# Big Bang 'Blow-Out'

Illustrated Science Exploration by Rolf A. F. Witzsche

© Text Copyright Rolf A. F. Witzsche 2018
all rights reserved

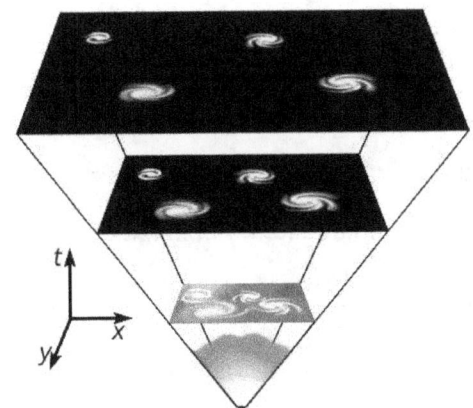

# Big Bang Blow-Out

This book contains the transcript with images of the exploration video with the above title:
see: http://www.ice-age-ahead-iaa.ca/

## Lead in:

The Big Bang theory of an explosively expanding and self-consuming entropic universe defies all visible evidence and reason. Its "Doppler Effect, Red-Shift theory" is self-evidently false, which blows away the prime postulate that the Big Bang theory is built on.

Let's celebrate that the Big Bang theory is false, because if it was true, humanity would have no hope. All of our presently used energy fuels are fast becoming depleted. At the present rate, oil and nuclear fuels, will be depleted in 60 years. Then what? What will power humanity for its future millions of years?

Fortunately, the universe is anti-entropic in nature. It is self-creating, self-powering, self-enriching, self-advancing. Nothing is winding down. Neither is our humanity winding down, but is self-advancing by the same principle. Humanity stands in defiance of entropy. It stands as the pinnacle of anti-entropic life, endowed with the scientific and technological capacity to tap into the electric energy streams that power the universe. With it, humanity has the power to create itself an infinite, energy-rich future, standing in defiance of the ice ages and of their consequences. This truth will assure our future past the point when the currently used energy fuels are depleted.

# Table of Contents

**Segment 1- Entropy versus Anti-Entropy** .................................................................... 14

    A grand theory of the origin of the universe ........................................................... 15

    The theory is built on a big mistake in perception ................................................... 16

    In the Big Bang cosmology ........................................................................................ 17

    Every sun in the universe .......................................................................................... 18

    Is the Universe Entropic? .......................................................................................... 19

    Entropic systems are common on Earth ................................................................... 20

    The basis for the entropic universe theory .............................................................. 21

    What is Red Shift? ..................................................................................................... 22

    When we measure red shift ...................................................................................... 23

    The Earth the center of the universe ........................................................................ 24

    The big ring of fire and the central void ................................................................... 25

    The entire Big Bang Creation theory ......................................................................... 26

    From the basis of plasma physics ............................................................................. 27

    The nature of the photon, the carrier of the light ................................................... 28

    A different principle applies to sound waves ........................................................... 29

    Light is a propagated stream of individual entities .................................................. 30

    Light consists of photons that are individual entities ............................................... 31

    How is the red-shift possible .................................................................................... 32

    A slight energy depletion occurs along the way ...................................................... 33

    Red-shift amounts do vary with local conditions ..................................................... 34

    The Big Bang theory is paradoxical .......................................................................... 35

    The Anti-Entropic Principle of the Universe ............................................................ 36

    This opposite of entropy ........................................................................................... 37

| | |
|---|---|
| David Bohm | 38 |
| The modern perception of sub-atomic particles | 39 |
| That's the mark of anti-entropy. | 40 |
| Every sun a creator of its surrounding worlds | 41 |
| The so-called 'Pillars of Creation' | 42 |
| Every sun fuses plasma into atoms | 43 |
| Anti-Entropic Energy | 44 |
| Celebrate that the Big Bang theory is false | 45 |
| Not a single sun in the universe is self-powered | 46 |
| The universe is powered on the cosmic scale | 47 |
| Our future depends on this utilization | 48 |
| Burning fuels for energy production | 49 |
| To tap into the cosmic electric energy streams | 50 |
| Plasma streams that are encircling the Earth | 51 |
| The Big Bang Without a Future | 52 |
| The Big Bang theory insists | 53 |
| When the energy resources are gone, our future ends | 54 |
| The concept by which everything ends | 55 |
| Even the Sun will die | 56 |
| Fortunately, we see no evidence | 57 |
| The observable universe is uniformly dense | 58 |
| We are seeing a universe that is actively self-powered | 59 |
| The Anti-Entropic, Self-Powered Universe | 60 |
| Back to David Bohm | 61 |
| A latent sea of water | 62 |
| Water vapor | 63 |
| The latent background of energy | 64 |

| | |
|---|---|
| The electron | 65 |
| A proton is made up of three quarks | 66 |
| The electron, in comparison | 67 |
| Quarks and leptons are energy in motion | 68 |
| Protons and electrons exist in space | 69 |
| Roughly 99.94% of the mass of the universe | 70 |
| Plasma, the lifeblood of the universe | 71 |
| The universe without plasma is inconceivable | 72 |
| All space is filled with plasma | 73 |
| Every atom that exists | 74 |
| The Big Bang creation theory | 75 |
| The creative process | 76 |
| Nothing ever had a single-point origin | 77 |
| The creative process is universal | 78 |
| Nuclear-fusion energy production | 79 |
| A sun is not entropic in nature | 80 |
| The universe is energy | 81 |
| We fail aiming for fusion power | 82 |
| Dead-end energy future | 83 |
| Explosion is entropic and blows itself out | 84 |
| By clinging to the Big Bang trap | 85 |
| The anti-entropic energy resource | 86 |
| If the Big Bang was the truth | 87 |
| Plasma flowing through a nebula | 88 |
| Our energy-future on Earth | 89 |
| The Red Square nebula | 90 |
| Our interface with the stellar plasma streams | 91 |

Also visible on the Sun ............................................................................................................. 92

The cosmic electric energy platform ......................................................................................... 93

The Big Bang Suicide Pact ....................................................................................................... 94

The Big Bang theory is a trap .................................................................................................. 95

The biofuels genocide contract ................................................................................................ 96

The consequences of the difference ......................................................................................... 97

In this trap, the Sun is deemed entropic .................................................................................. 98

This dying-star solar model ..................................................................................................... 99

For as long as the Big Bang cosmology rules .......................................................................... 100

The Plasma Universe offers itself as an open door ................................................................. 101

The Plasma Universe ............................................................................................................. 102

Segment 2 - Entropy - Empire – Economic ............................................................................... 104

The Big Bang trap is paraded as brilliant ................................................................................ 105

With the Big Bang theory ...................................................................................................... 106

The lyrics for the empire-song are a lie ................................................................................. 107

While no empire has ever survived ......................................................................................... 108

Society has freed itself and its future ..................................................................................... 109

By placing itself onto the platform of anti-entropy .................................................................. 110

On this platform nothing is winding down .............................................................................. 111

Empire lacks the intention to be creative ............................................................................... 112

The Big Bang theory needs to be dealt with ........................................................................... 113

Theory of the inner emptiness of the universe ....................................................................... 114

The Big Bang philosophy ....................................................................................................... 115

Humanity becomes free ......................................................................................................... 116

The principle of cosmic anti-entropy ...................................................................................... 117

The Big Bang theory blocks humanity .................................................................................... 118

The plasma universe concept ................................................................................................. 119

Big Bang theory to prevent the collapse of empire ... 120

Empire projected its death-model onto the universe ... 121

Empire is an empty hole that drains the world ... 122

The current world-empire is no exception ... 123

Empire is worse than just being empty ... 124

The Big Bang concept is choking science ... 125

Anti-Entropy in Civilization ... 126

The human bang, the real big bang ... 127

We can get to this stage of freedom ... 128

In the shadow of blocking of humanity ... 129

Our present stage is precarious ... 130

Running away from anti-entropic economics ... 131

Not a platform for civilization ... 132

Two Opposing Platforms in Economics ... 133

The anti-entropic platform ... 134

A bridge across the tropics ... 135

Credit to get an enriching job done ... 136

Economics in civilization ... 137

The money-bag system ... 138

Empire is doomed by its own premise ... 139

The consequences are world-destructive ... 140

The anti-entropic system of economics ... 141

We are the supreme being on Earth ... 143

Segment 3 - "Lord of the Rings" & The American Paradox ... 144

Humanity is anti-entropic in nature ... 145

Society's illusions about the universe ... 146

In the Big Bang dream ... 147

Whether humanity achieves the rich future ............................................................................. 148

The kind of choice that Tolkien places before society .............................................................. 149

Unfortunately our world is darker ............................................................................................ 150

Our civilization is choked with false theories .......................................................................... 151

No truth in any of these theories .............................................................................................. 152

War is not a natural element of humanity ................................................................................ 153

Monetarist thievery has never created a nobler and stronger society ....................................... 154

limited nuclear war' is a false theory ....................................................................................... 155

Extinction is the ultimate of entropy ........................................................................................ 156

From entropy to anti-entropy .................................................................................................... 157

The Pioneering Vision in The Lord of the Rings ..................................................................... 158

Tolkien takes us on a fictional journey .................................................................................... 159

For the Lord of the Rings ......................................................................................................... 160

Lesser rings, rings of gold ........................................................................................................ 161

Casting the theory of entropy into oblivion ............................................................................. 162

Tolkien leaves no room for a compromise ............................................................................... 163

The "Glass Steagall" banking legislation ................................................................................. 164

Among the foes in the Lord of the Rings ................................................................................. 165

We have many evil potentates in high places .......................................................................... 166

The supernova war that nothing survives ................................................................................. 167

Masters of false theories ........................................................................................................... 168

Russia and China placed at the cross-hairs .............................................................................. 169

Where people lie to themselves ................................................................................................ 170

When society looses its renaissance of the truth ...................................................................... 171

The American Paradox ............................................................................................................. 172

America stands as a paradox .................................................................................................... 173

Why has the nation fallen ......................................................................................................... 174

- The universe doesn't fall into entropy .................................................................................... 175
- America the paradox of a fallen star ..................................................................................... 176
- The paradox that America became ....................................................................................... 177
- One of the great masters of the trap ..................................................................................... 178
- Adam Smith was mistaken ..................................................................................................... 179
- When money is an object for stealing ................................................................................... 180
- In the extreme case, as we have it today .............................................................................. 181
- Nuclear war, in any form ........................................................................................................ 182
- A dead peace without a human voice ................................................................................... 183
- Far distant from Adam Smith's economics ........................................................................... 184
- The future of humanity ........................................................................................................... 185
- Real economics aims for the goodness in living ................................................................. 186
- After India became a free nation in 1947 ............................................................................. 187

Segment 4 - Colonial Age & Glass Steagall Compromise ..................................................... 188
- The colonial rule for stealing the wealth of nations ............................................................. 189
- The Colonial Age, .................................................................................................................... 190
- To circumvent the inherent entropic collapse ...................................................................... 191
- The colonial age began .......................................................................................................... 192
- When the resistance was internal ......................................................................................... 193
- Almost the entire world became subjugated ....................................................................... 194
- Lists of the subjugated nations ............................................................................................. 195
- The American republic was born ........................................................................................... 196
- America did attain its freedom, and fought to retain it ....................................................... 197
- America lost itself again to the devil within ......................................................................... 198
- Colonial wars soon became world wars ............................................................................... 199
- Stealing with the force of 'invincible' arms .......................................................................... 200
- Few soldiers in history knew ................................................................................................. 201

War has become a worldwide disease .................................................................................................. 202

Stealing by all means possible, humanity is doomed ...................................................................... 203

As stealing demands evermore wars ................................................................................................. 204

The whole of humanity is now doomed ........................................................................................... 205

Their greed demands, a scientific excuse ......................................................................................... 206

Science complied................................................................................................................................... 207

Thievery inherent in the kingdom of empire .................................................................................. 208

A new wind is rising in the distant lands......................................................................................... 209

Where America had stood when it stood tall ................................................................................. 210

The Glass Steagall Compromise ........................................................................................................ 211

With the repeal of the Glass Steagall Act......................................................................................... 212

The new colonialism of the Euro empire ......................................................................................... 213

The freedom to steal has become protected .................................................................................... 214

War against Russia and China ........................................................................................................... 215

In America, the Glass Steagall legislation was repealed in 1999 ................................................. 216

Compromise on principle became its doom ................................................................................... 217

Entropy and Anti-entropy are mutually exclusive ........................................................................ 218

The theory of self-consuming stars................................................................................................... 219

We play the same compromising game ........................................................................................... 220

The Glass Steagall compromise......................................................................................................... 221

Twenty years before Roosevelt .......................................................................................................... 222

The year 1999 marks the historic beginning.................................................................................... 223

To restore the Glass Steagall law ....................................................................................................... 225

Solving the tragedy at this stage ....................................................................................................... 226

To merely reinstate Glass Steagall, defies the nature of reason ................................................... 227

Meeting the Ice Age Challenge ......................................................................................................... 228

To create a more-just economic order in America ......................................................................... 229

- Bring the future demands into the present .................................................. 230

# Segment 5 - The Need for Looking Forward ............................................. 231

- Nothing is gained from clinging to the past ................................................ 232
- By becoming latched to the past .............................................................. 233
- By staying latched to the past .................................................................. 234
- Without the advancing recognition in society ............................................. 235
- Glass Steagall has become too shallow ..................................................... 236
- The imperative comes with the Ice Age Challenge ..................................... 237
- It challenges us all to become human beings ............................................ 238
- To create an Ice Age Renaissance ............................................................ 239
- We need to look forward with the eyes of science ..................................... 240
- In terms of our natural capacity as human beings ..................................... 241
- The Ice Age phenomenon is not the product of entropy ............................. 242
- Children of the anti-entropy of the universe .............................................. 243
- Ice ages are critical elements in the progressive dynamics ........................ 244
- The Big Bang theory stands plainly in denial ............................................. 245
- Big Bang theory might have been intentionally staged .............................. 246
- The Big Bang Cosmology .......................................................................... 247
- The Ice Age is near, as near as the 2050s ................................................. 248
- A great need for a renaissance of truth ..................................................... 249
- The proof of the 'pudding' is unmistakable ................................................ 250
- the sky is no limit" or "there are no limits." ............................................... 251
- With increased industrialization ................................................................ 252
- Population increase mirrors anti-entropic economics ................................ 253
- The rate of increase that can be achieved ................................................. 254
- Floating bridges across the tropical seas .................................................. 255
- The 6000 new cities for a million people each .......................................... 256

To break down the barrier in the mind ... 257

Drought conditions can be offset ... 258

Desalination of ocean water provides an endless resource ... 259

Recognized already during the Kennedy era ... 260

Deep ocean reverse osmosis desalination ... 261

Deep ocean reverse osmosis desalination ... 262

India can never suffer water shortages ... 263

The relocation of entire nations becomes essential ... 264

We will built the 6000 new cities ... 265

Whether we survive the Ice Age Challenge ... 266

We have ample of proof to our credit ... 267

Entropic factions in politics and philosophy ... 268

As the truth is being experienced, we begin to fly high ... 269

More Illustrated Science Books by Rolf A. F. Witzsche ... 270

# Segment 1- Entropy versus Anti-Entropy

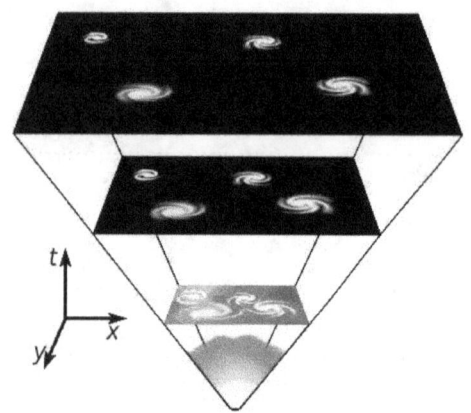

# Big Bang Blow-Out

**Segment 1    Entropy versus Anti-Entropy**

Segment 1- Entropy versus Anti-Entropy
Segment 2 - Entropy - Empire - Economic
Segment 3 - "Lord of the Rings" & The American Paradox
Segment 4 - Colonial Age & Glass Steagall Compromise
Segment 5 - The Need for Looking Forward

# A grand theory of the origin of the universe

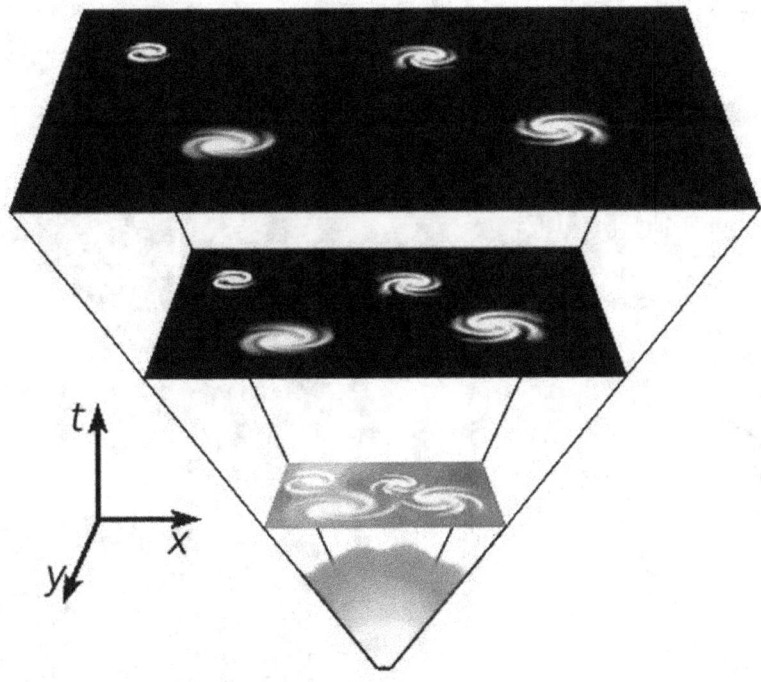

The Big Bang cosmology is built on a grand theory of the origin of the universe.

It's story begins with the postulate that nothing existed before the Big Bang 13.8 billion years ago. It is also said that whatever there might have been, collapsed into itself and sparked a giant explosion in which all the atoms in the entire universe were formed, from which, subsequently, the universe was born. It was born by condensation of the created primordial dust that is deemed to be expanding at ever-greater speed, raising away from the point of explosion.

It is said, that as the expansion 'progressed,' the speeding dust condensed into stars and planets and galaxies, which will remain active for a season, towards their eventual heat death when all the energy from the Big Bang is consumed.

## The theory is built on a big mistake in perception

"Is the Universe Entropic?"
"What is Red Shift?"
"The Anti-Entropic Principle of the Universe"
"Anti-Entropic Energy: Electric Power Forever"
"The Big Bang Without a Future"
"The Anti-Entropic, Self-Powered Universe"
"The Big Bang Suicide Pact"
"The Big Bang: A Lie Projected onto the Universe"
"Two Opposing Platforms in Economics"
"The Pioneering Vision in the Lord of the Rings."
"The American Paradox"
"The Colonial Age, the Age of 'Fierce' Entropy"
"The Glass Steagall Compromise"
"The Need for Looking Forward"

The Big Bang theory and its effects

a study in 14 parts

The video focuses on a big subject that calls for a big production. The subject is that big, however, it is big only because the theory is built on a big mistake in perception that has affected the shape of civilization more deeply, on a wider front, and for a longer period of time, than any other misperception in the history of humanity.

The mistake in perception actually precedes the Big Bang theory by several millennia, with numerous related effects old and new that are often deemed unrelated, but which, when they are brought together into a single complex, tell a story that cannot be easily recognized otherwise. For this reason, the video became somewhat lengthy. And yes, the Big Bang theory, as a concept of the cosmos, stands in the middle of it all by its role in the larger historic context, in ways that are not apparent when the theory is looked upon, standing in isolation.

To begin, let's look at the theory itself, and the story it tells.

## In the Big Bang cosmology

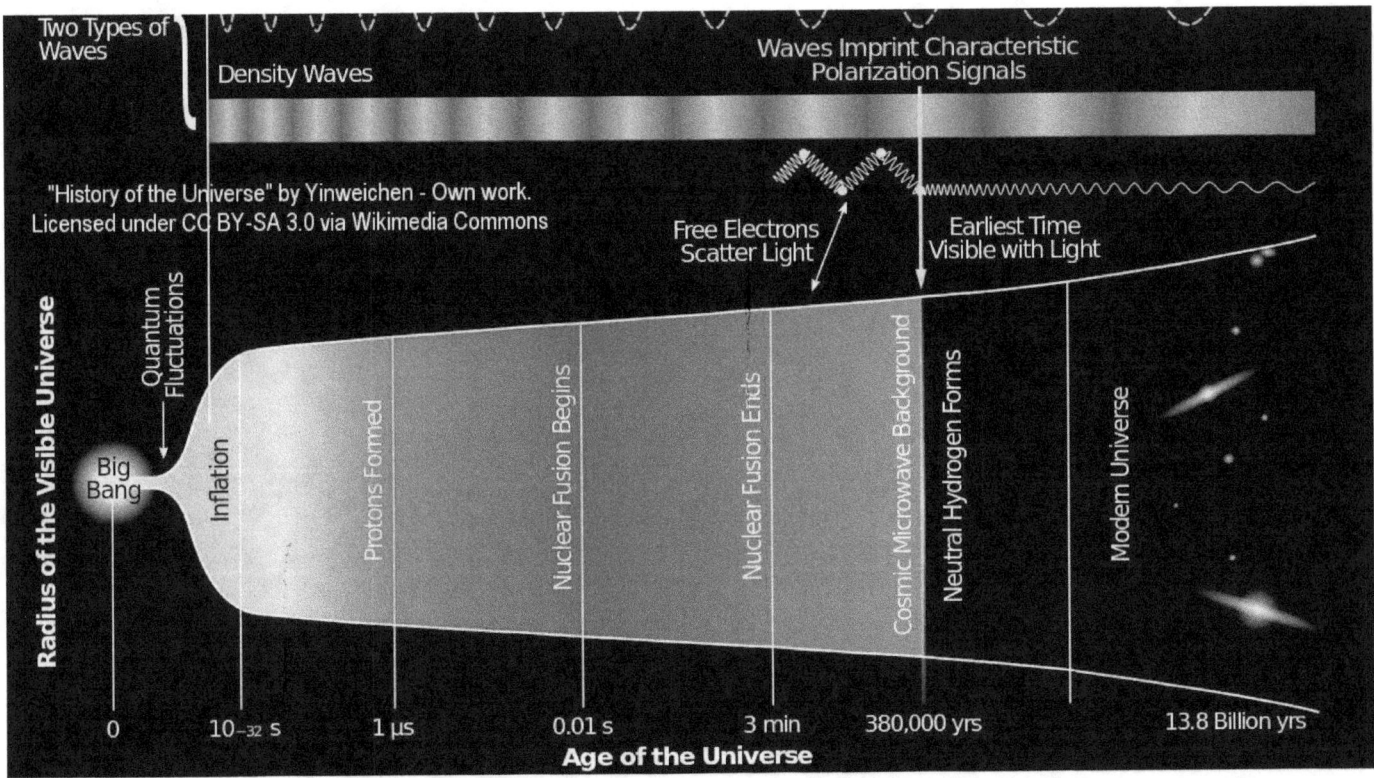

In the Big Bang cosmology the universe is regarded to be inherently entropic in nature. All the energy and substance of the universe is deemed to have originated at a single-point source, at a single moment, as a single package. From this point on, it had all the energy and substance that it will ever have. Everything thereafter is said to have expanded outward, but with a built-in process of decay towards its death by energy depletion. The theory thereby presents the universe as entropic in nature. It is deemed to be self-consuming, winding down, and consequently self-collapsing. The term, entropy, defines an inherently collapsing dynamic system that is winding itself down by a process of depletion.

# Every sun in the universe

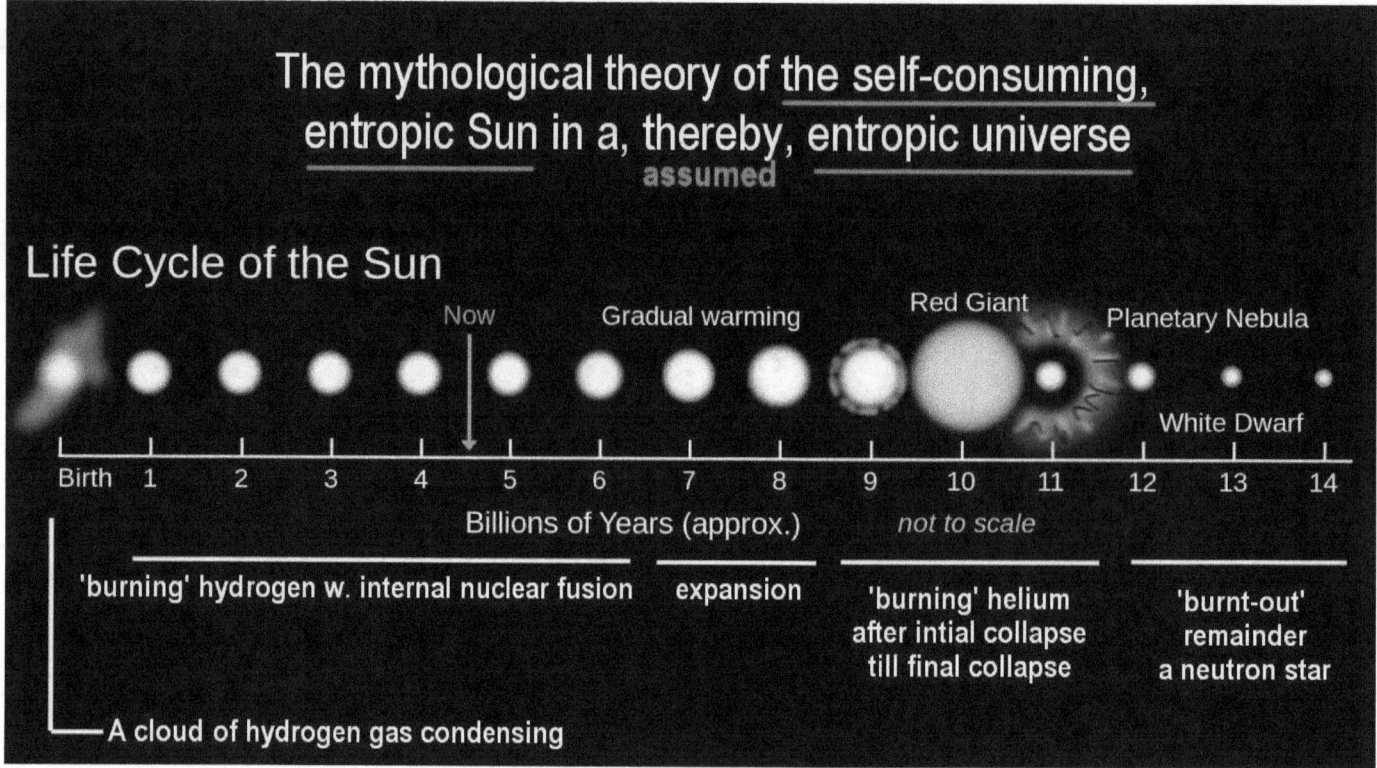

Every sun in the universe is 'seen' in this context. It is believed to be self-consuming as it burns the hydrogen gas that it is made of. The hydrogen gas is deemed to power a process of nuclear fusion deep within a sun, which is said converts hydrogen into helium with the release of free energy. By this theory, after a few dozen billion years have passed, every sun in the universe will have burned itself out, towards the end of the universe itself. The death of the universe is thereby deemed to be assured by the built-in entropy of the cosmic system that is simply unavoidable. Can this be true?

## Is the Universe Entropic?

"Is the Universe Entropic?"

## Entropic systems are common on Earth

Entropic systems are common on Earth. We are familiar with the concept of entropy. We find entropic systems in the form of clocks, watches, and wind-up toys that run down and stop when the energy in their spring or their battery is spent. That's the effect of entropy. The Big Bang theory tells us that the universe operates in the same manner, though much more slowly. It is said to have expanded from an intense single-point explosion 13.8 billion years ago, that wound it up, that got it going, that got it expanding further and further until it fades into nothing when the energy that came with the explosion, is used up. Can this be true?

# The basis for the entropic universe theory

The center of the Milky Way, at the center of the Big Bang explosion of the universe

The basis for the entropic universe theory, which the Big Bang theory may have been derived from, is the red-shift effect of light coming from distant galaxies. The more distant the observed galaxies are located, the greater is the red shift that is being observed in the light received from them. The stated theory is, that the red-shift is caused by the observed objects receding away from us. This simply means, that the greater the red-shift is, of the light from observed galaxies, the faster the galaxies are speeding away from us, the observer. But here a paradox begins to unfold that unravels the Big Bang theory.

The paradox is that the red-shift in light is observed in all directions. This means that the Earth is once again believed to be the center of the universe, as had been believed in medieval times, so that the entire universe is deemed to be speeding away from us. Wow! But is this true? Does this make sense?

In whichever direction we look, various amounts of red-shift are observed, while not the faintest blue-shift is ever observed. This evidence makes one suspicious, doesn't it?

The measured red shift is evidently real, but by it being real, it places the most fundamental platform of the Big Bang theory into doubt, and everything with it that it is built on it, because it simply makes no sense that miraculously the Earth should be the center of the universe with everything speeding away from us in all directions. The proposition places the entire theory of the entropic nature of the universe seriously into doubt.

## What is Red Shift?

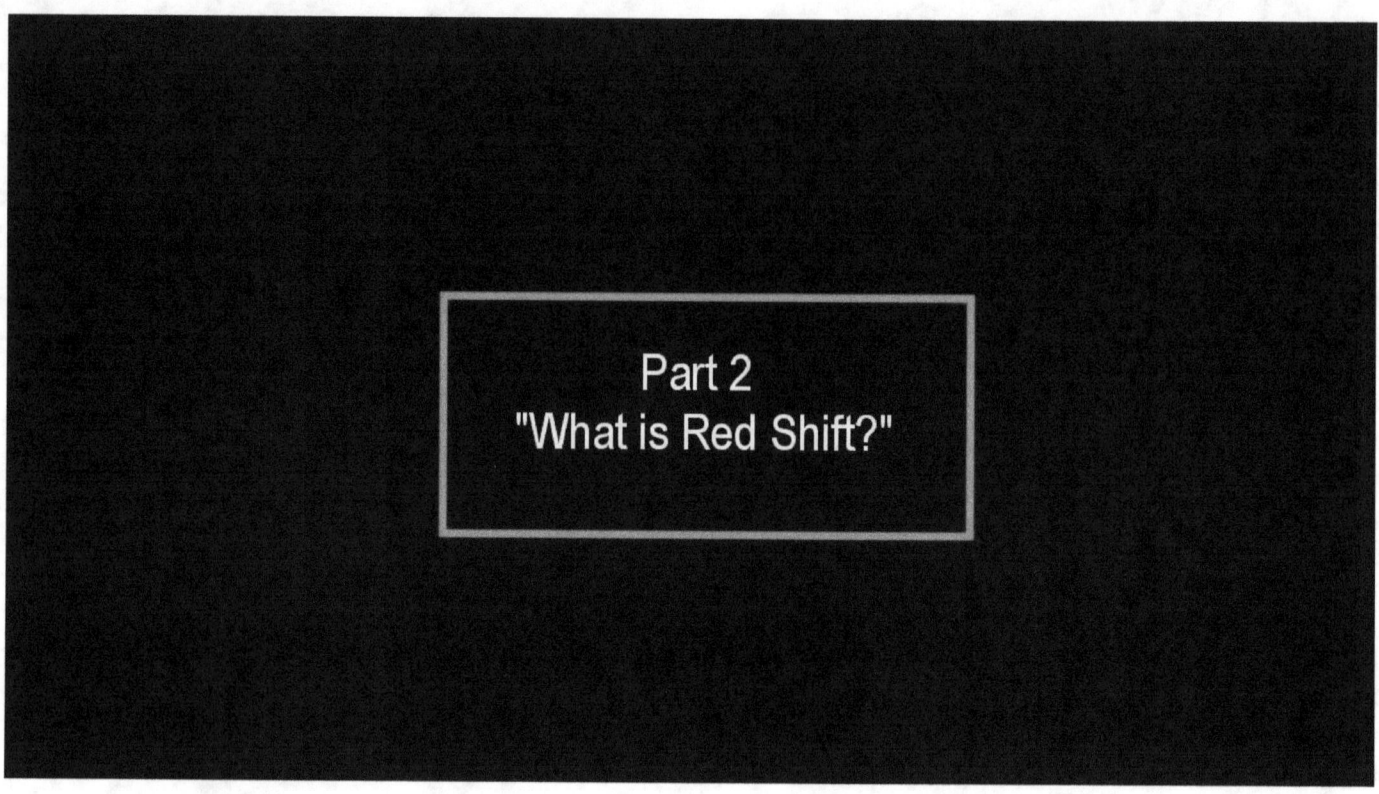

"What is Red Shift?"

# When we measure red shift

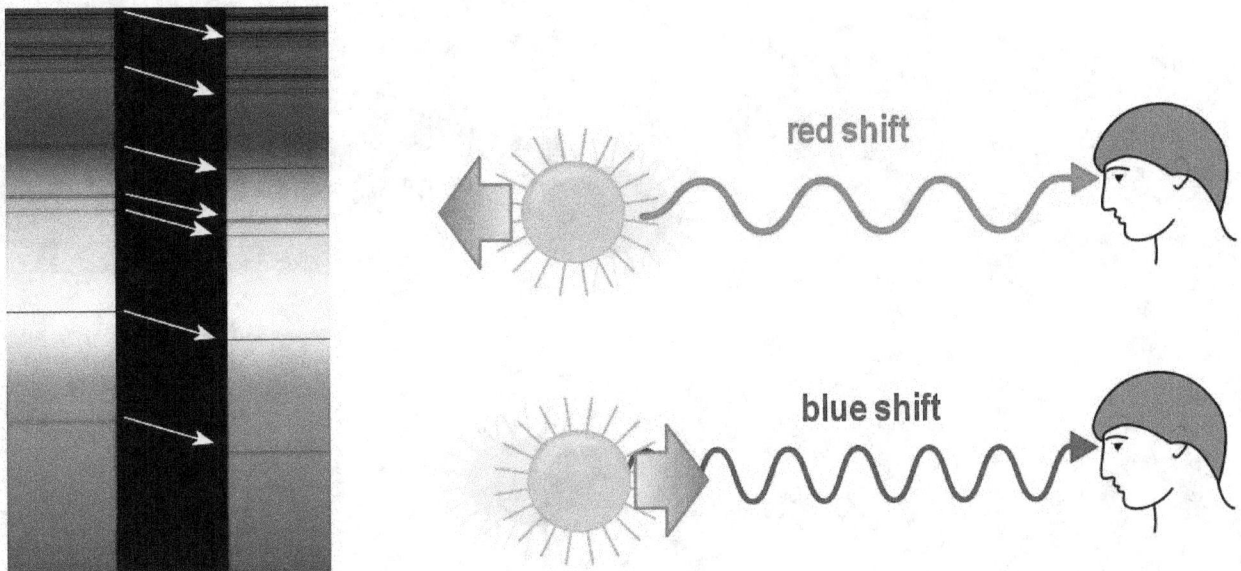

the red-shift of light from distant galaxies is assumed to indicate that the galaxies are fast moving away - stretching the light

What do we measure when we measure red shift?

It has been theorized that when a source of light is moving away from an observer, the light-waves are stretched out, which causes the observer to perceive the light shifted towards longer wavelengths: towards the red of the visible spectrum.

Inversely, it has been theorized that when a light source is fast approaching an observer, the light is compressed by the movement towards shorter wavelengths, which is deemed to cause a shift to occur, towards the blue.

The shifting itself is observed in the shifting absorption lines in the light spectrum. The individual lines are caused by specific colors of light being absorbed by specific atomic elements in the path of the light. The resulting absorption pattern is known to be essentially uniform among the galaxies, only the observed shifting of the pattern varies. From the amount of the shifting, and from the resulting calculated expansion speed of the most distant galaxys, it has been theorized that the Big Bang creation event occurred 13.8 billion years ago. But is the theory true? Can it be true?

# The Earth the center of the universe

The center of the Milky Way, at the center of the Big Bang explosion of the universe

Is the Earth really the center of the universe, with all galaxies racing away from us? Does the wide field of evidence that we see, match the assumption?

No, it really doesn't. The wonderful Big Bang tale appears to be full of self-evident holes. One of the biggest of these holes is the hole that doesn't exist.

## The big ring of fire and the central void

When a giant explosion occurs in space, the explosion creates a thin shell of fire around what fast becomes an empty center. The result would be similar to what we see here. But this is not what we see through the telescopes, when we look at the universe. The big ring of fire and the central void that we should see, don't exist.

## The entire Big Bang Creation theory

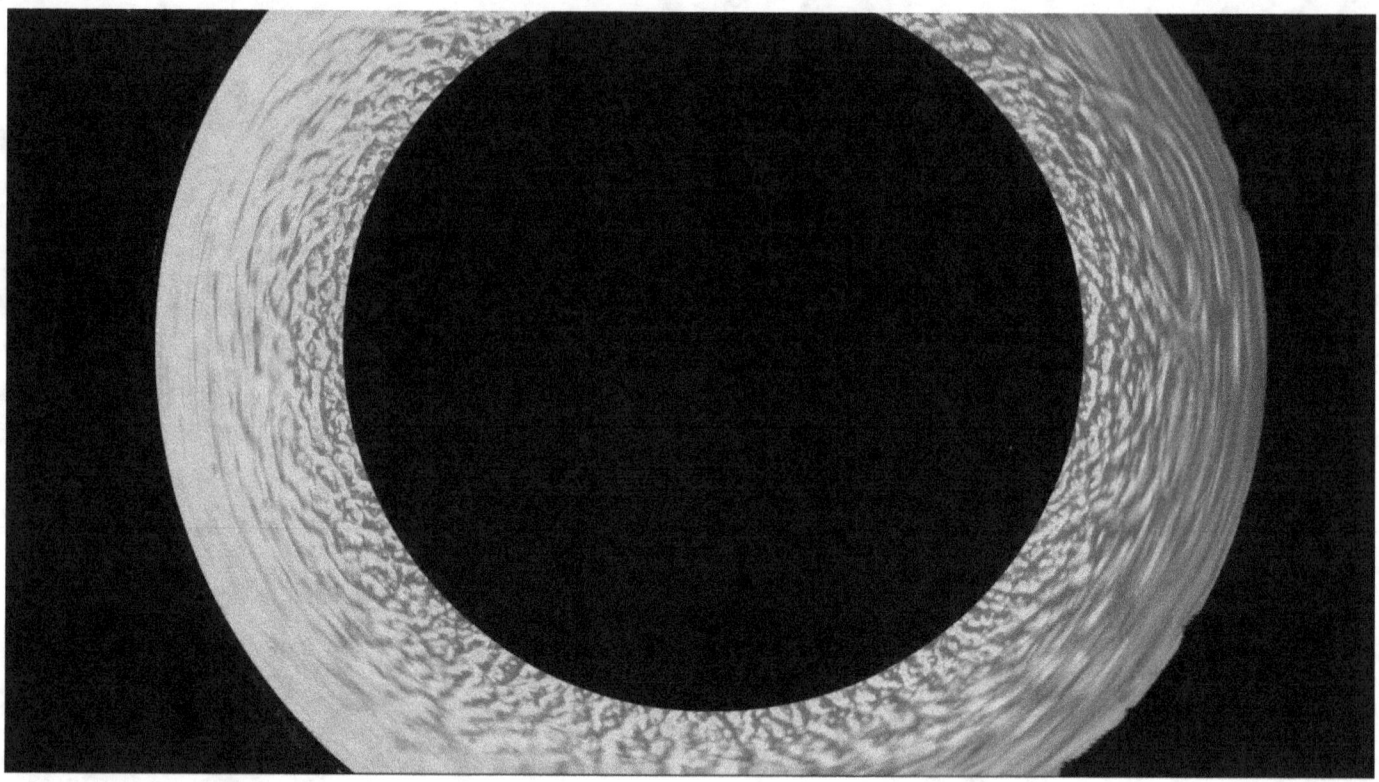

No central void has ever been found. No ring of fire with an empty center has been recognized. So what about the red-shift then? What happens to the light from distant galaxies, especially from those at the outer edge that are deemed to be racing away from us? What happens to the light that causes it to red-shift?

That's in important question, because the entire Big Bang Creation theory is fundamentally built on a specific assumption for the red shift phenomenon? How does one sort out the truth?

# From the basis of plasma physics

The center of the Milky Way, at the center of the Big Bang explosion of the universe

When we look at the universe from the basis of plasma physics, the red-shift phenomenon, and the paradox of the missing evidence for the Big Bang, become rather easily solved. The key for this is found in the unique nature of light.

# The nature of the photon, the carrier of the light

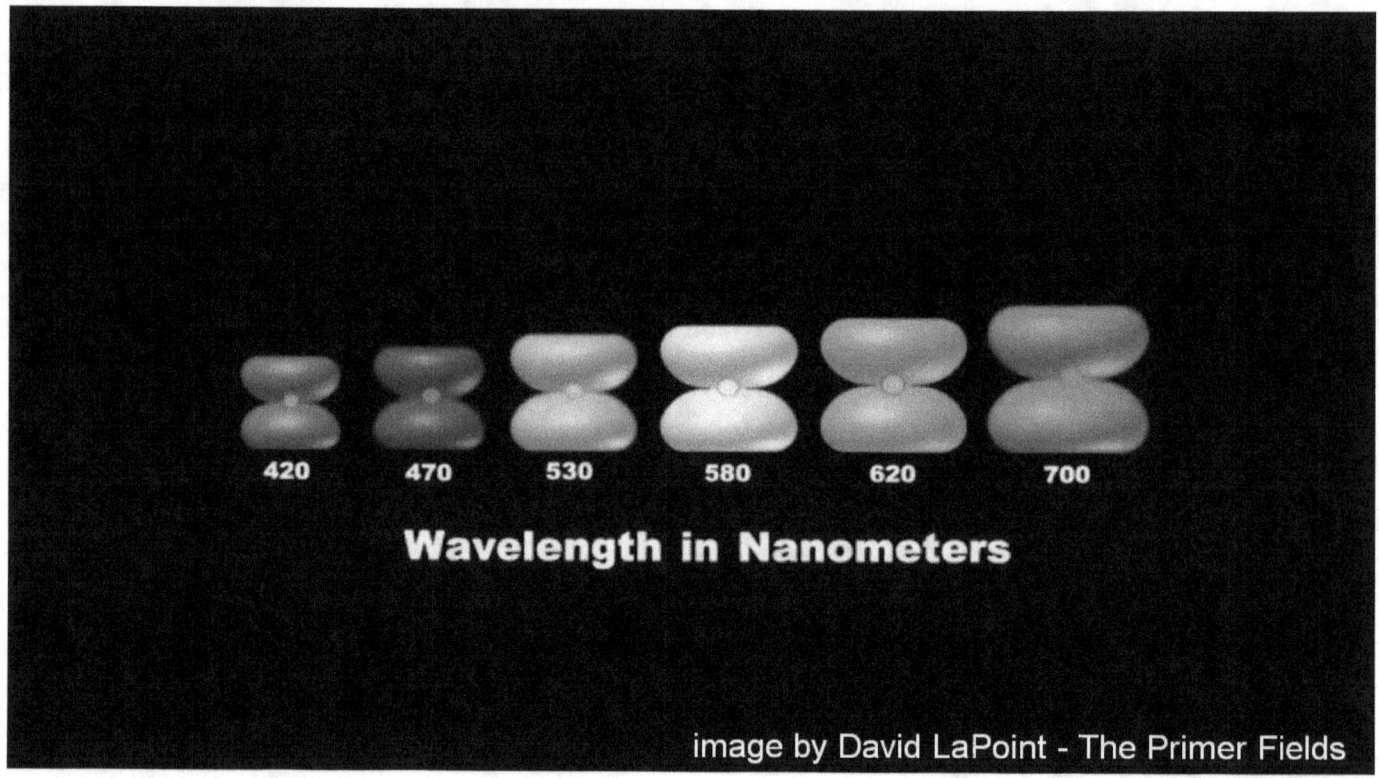

image by David LaPoint - The Primer Fields

The nature of the photon, the carrier of the light, is such that a fast moving light source doesn't actually alter the 'color' of the light. The fact is easily recognizable.

The color of light is determined by the energy environment within the electromagnetic photon envelope, when the light is created. Some photon packets are larger, some are smaller. The higher the energy level is, inside a photon package, the tighter the package is held together, and the smaller it thereby becomes. The physical movement of an atomic element in which a photon package is formed, does not affect the size of the package. Only the energy-level within the atom affects the size of the photon and thereby its color. Different size photons are recognized as different colors. This fundamental pattern, of course, extends far beyond the visible spectrum, which is actually quite narrow. Extremely high energy levels, for example, produce extremely tiny packets, such as the x-ray 'photons' that are typically 100,000 times smaller than the photons of visible light. But why won't light change its color, once it is created? Let's take a look at that.

# A different principle applies to sound waves

The Tokaido Shinkansen high-speed line in Japan - Wikipedia

When one throws an apple out of the window of a fast-moving train, the apple doesn't end up being stretched out before it impacts the ground. This is so, because an apple is an entity that remains as it is, regardless of the speed of the train that carries it. However, a different principle applies to sound waves.

## Light is a propagated stream of individual entities

"Gare de Lyon TGV orange" by Smiley.toerist - Own work. Licensed under CC BY-SA 3.0 via Wikimedia Commons

When a fast train drives through a station, its whistle is of a higher pitch when it approaches the station, and of a lower pitch when it moves away from the station. This is so, because a sound wave is a disturbance in the air that reaches an observer's ear at a different rate by approaching or departing. In sound, the shifting pitch is normal, because sound is a disturbance in the air, while light is not a disturbance, but is an individual object, a discrete package. A beam of light is a propagated stream of individual entities.

# Light consists of photons that are individual entities

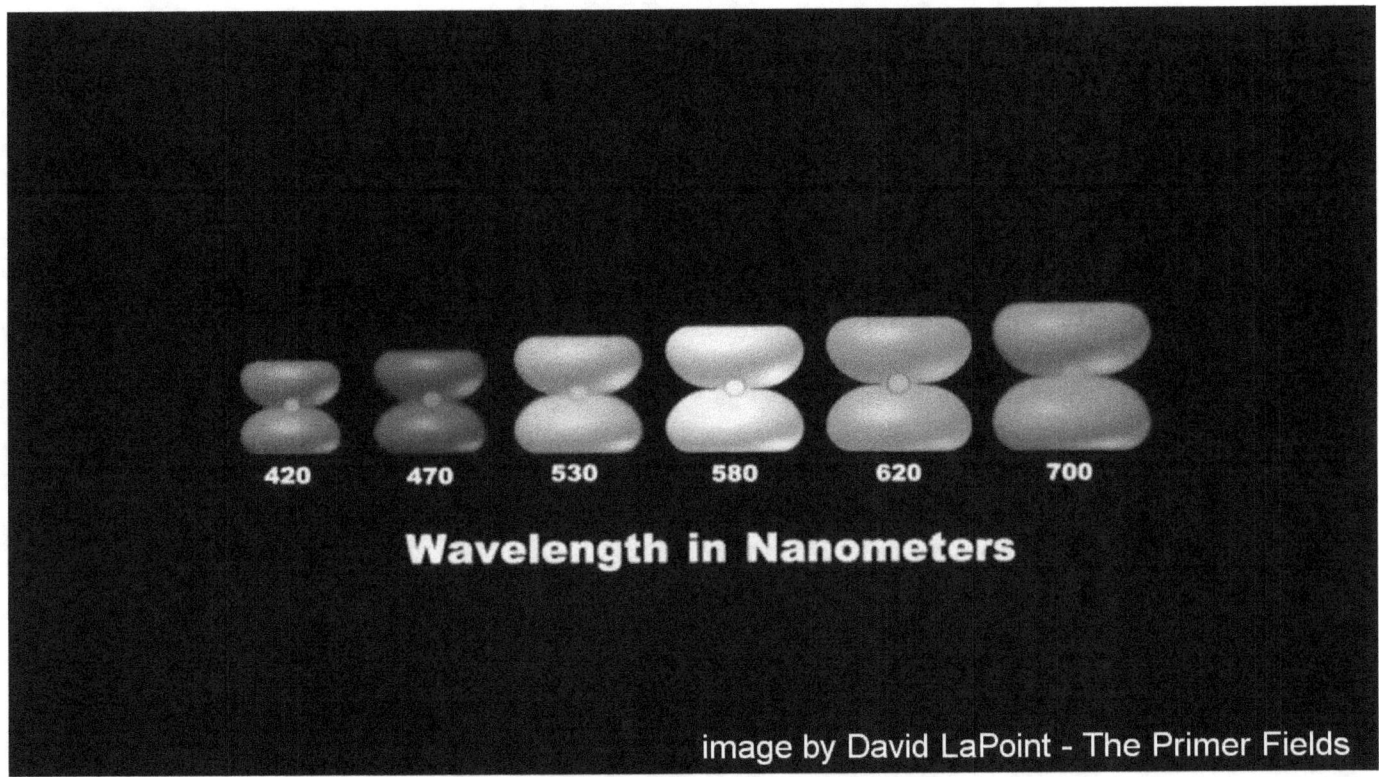

image by David LaPoint - The Primer Fields

However, because light consists of photons that are complete individual entities that are of a specific size for a specific color, and the size of the photon is determined by the energy it contains, the size can change when its energy is dissipated on the path of its travel over extremely long distances.

# How is the red-shift possible

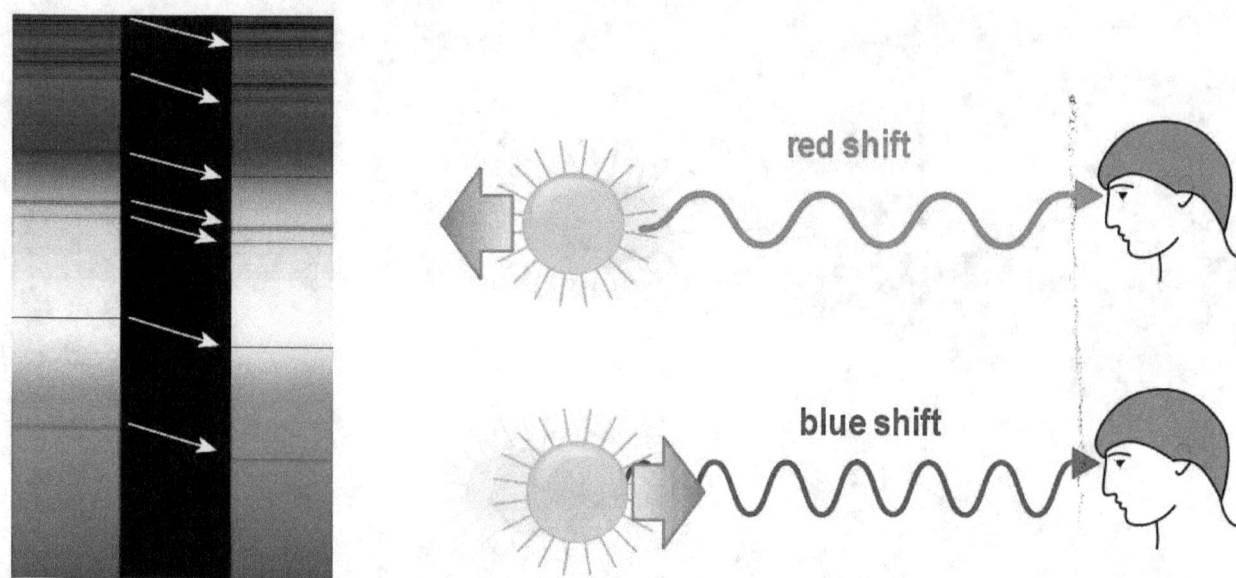

**the red-shift of light from distant galaxies is assumed to indicate that the galaxies are fast moving away - stretching the light**

This factor is significant in the context of the the Big Bang red-shift theory. The red shift that is observed in light from distant galaxies is, of course, totally real. We see the spectral lines received from distant galaxies. We see them shifted collectively towards the red.

But how is the red-shift possible when light is made up of discrete photon entities that cannot be stretched in the same manner as a disturbance would be stretched by a receding source?

## A slight energy depletion occurs along the way

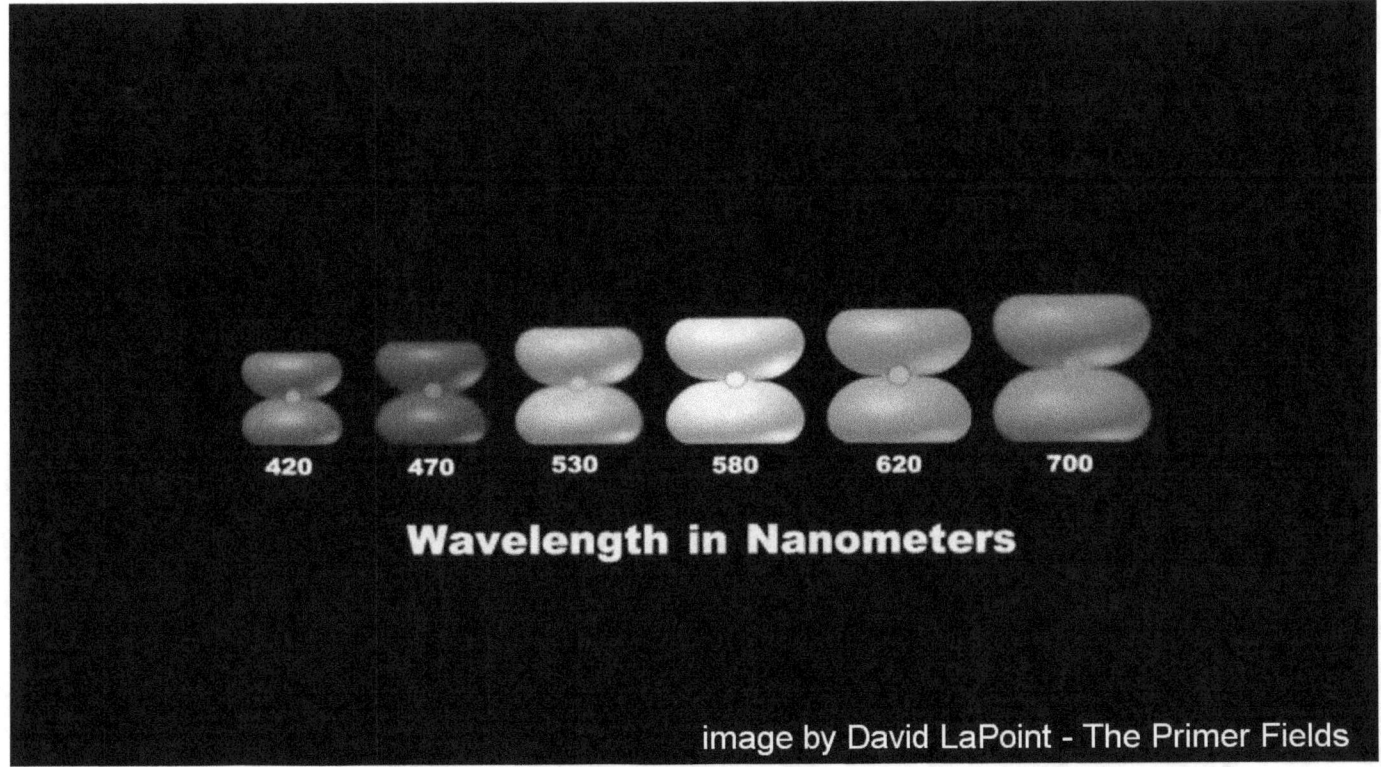

image by David LaPoint - The Primer Fields

The answer is simple. When light is propagated over very long distances in the range of hundreds of millions of light years, a slight energy depletion occurs along the way. The photon package increases in size by the energy depletion. The purple light thereby becomes blue light, it becomes a larger package, and the blue light becomes green light, and the green light becomes yellow light, and so on. The entire spectrum of the light becomes shifted towards the red, and the red, of course, gets shifted off the visible spectrum.

Now, the red shift phenomenon makes sense. It has nothing to do with light from distant galaxies being stretched by a moving light source that is racing away from us. The entire Big Bang theory, thereby falls apart.

Yes, the wavelength of light appears to be stretched thereby, but this is the result of the changing size of photons that results from gradual energy depletion.

The amount of the red-shift is typically a factor of the distance that light has to cross from far-away galaxies before it arrives at our door.

# Red-shift amounts do vary with local conditions

The center of the Milky Way, at the center of the Big Bang explosion of the universe

The amount of the red-shift is also a factor of the density of the cosmic dust, gases, and plasma that the light encounters along the way. This too, affects the rate of energy depletion.

For this reason, the red-shift amounts do vary with local conditions, both at the source, and along the way. The red shift definitely is not an indication that galaxies are racing away from us in all directions, as if we were at the center of the Big Bang universe.

# The Big Bang theory is paradoxical

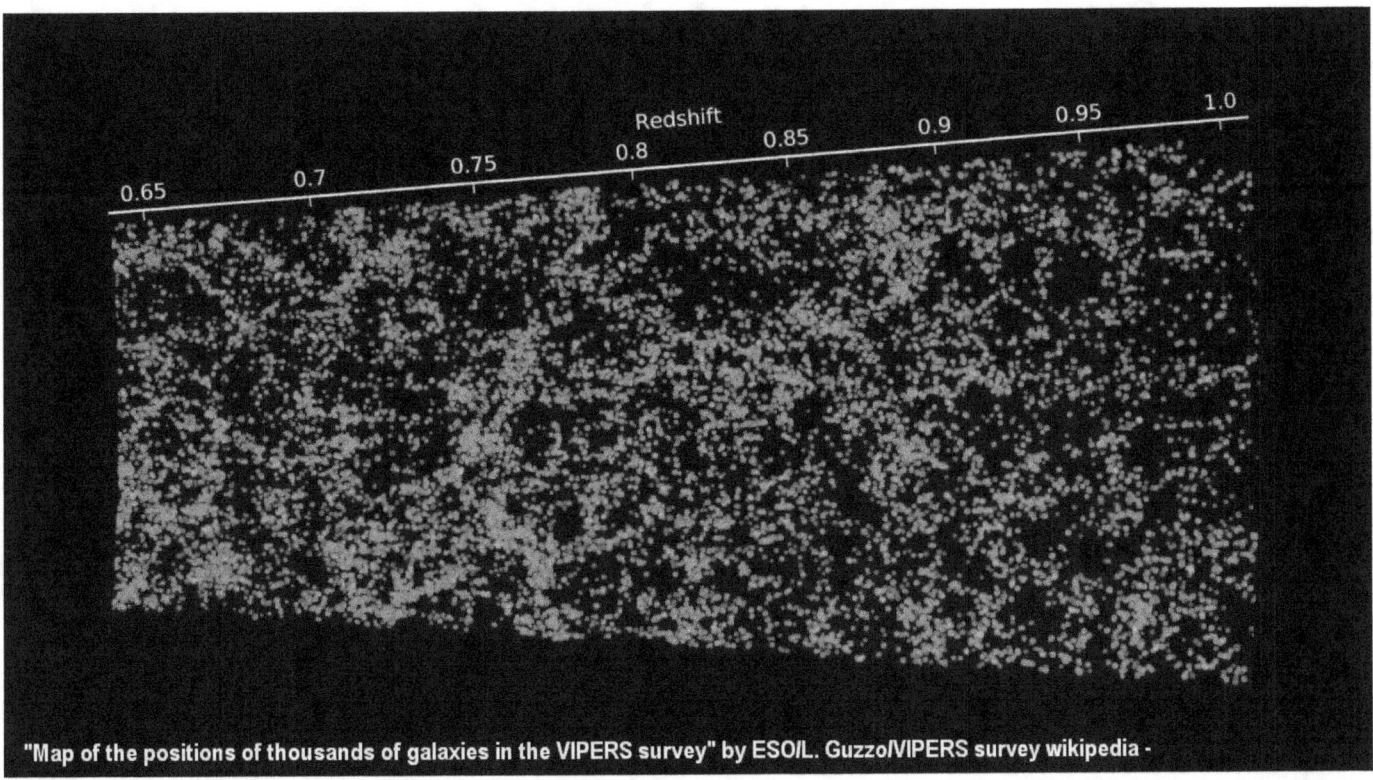

"Map of the positions of thousands of galaxies in the VIPERS survey" by ESO/L. Guzzo/VIPERS survey wikipedia -

The Big Bang theory is paradoxical for also another reason. This too, is rather obvious. The theory states that all matter in the universe was created in one place in the first 3 minutes of the explosion, and has been expanding outwardly thereafter for 13.8 billion years. If this was true, the material density of the universe would diminish outward with the cube of the distance from the source. At the rate of expansion that is theorized, the material distribution in distant regions should be so thinly spread that almost nothing should exist in distant places. Instead, the opposite is true.

# The Anti-Entropic Principle of the Universe

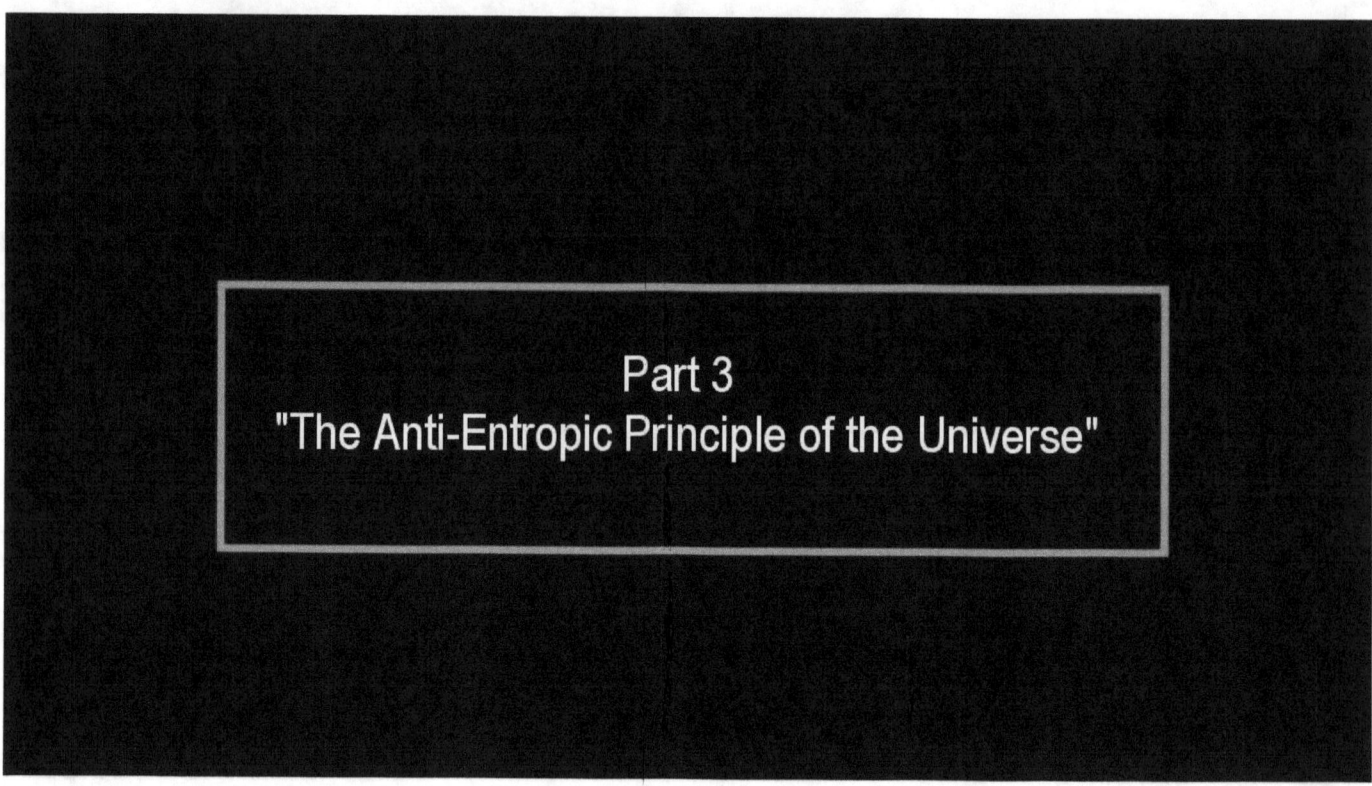

"The Anti-Entropic Principle of the Universe"

# This opposite of entropy

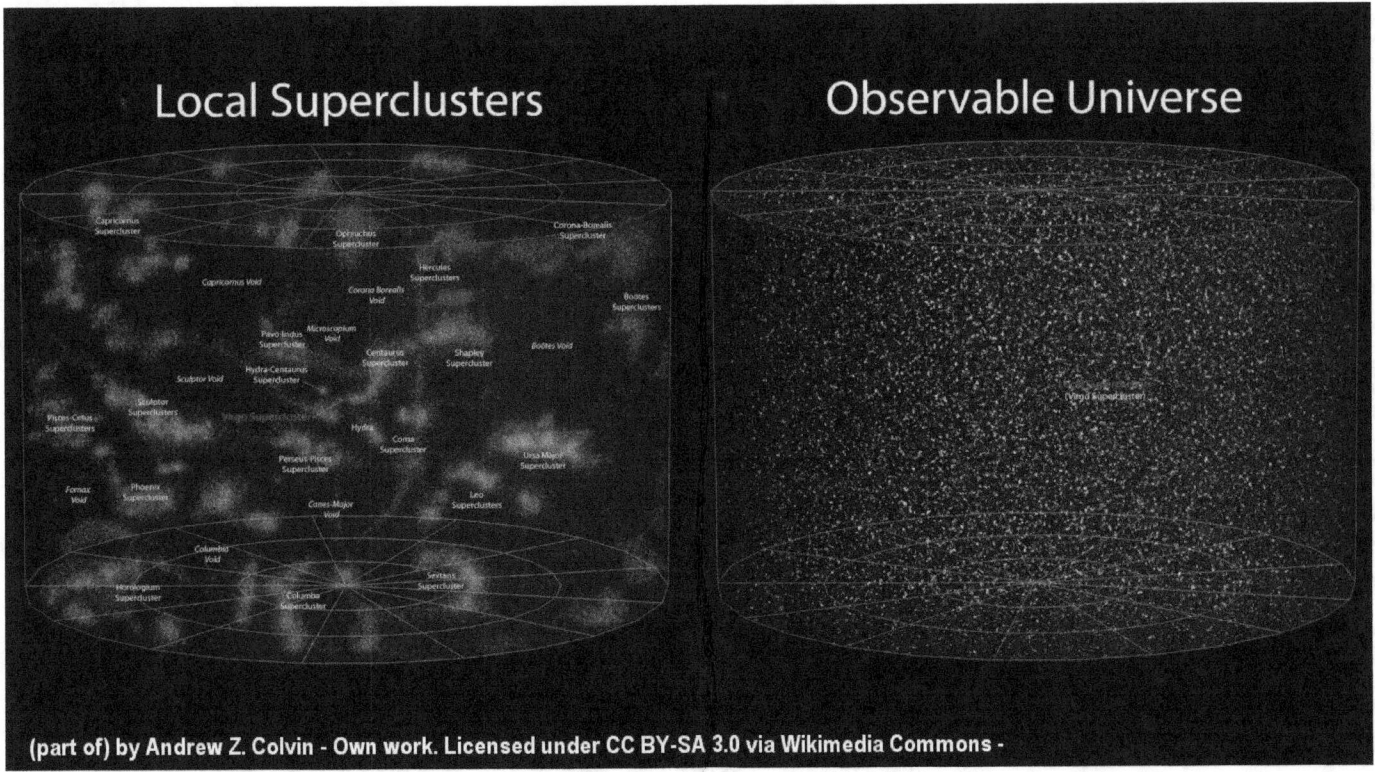

We see a universe that is of near homogenous density in populations of galaxies, clusters of galaxies, and super clusters. This evenly packed universe that we see was evidently not created from a single source at a single time, but is a universe that manifests itself as a self-unfolding plasma universe that is formed by creative principles that are manifest everywhere throughout cosmic space. The evidence suggests that the universe is not entropic in nature and winding down from its initial infusion of energy, but is completely self-creating everywhere, unfolding itself in a process that is opposite in nature to entropy. This opposite of entropy, one might term, anti-entropy.

# David Bohm

David Bohm, whom Einstein is said to have called his successor, speaks of cosmic space as latent energy that has an implicate order and an explicate order.

# The modern perception of sub-atomic particles

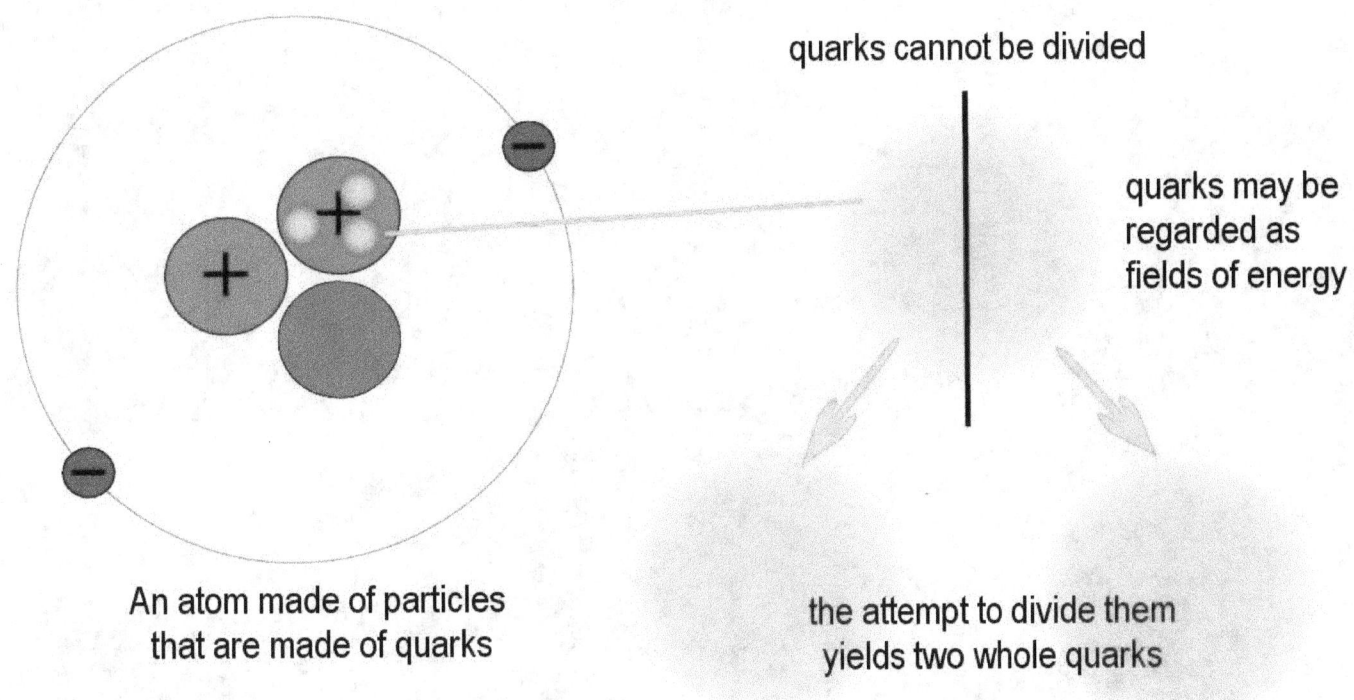

With the modern perception of sub-atomic particles, such as quarks, as being made up of moving points of energy, which subsequently make up the protons and electrons of the universe, it can be said that everything in the universe has ultimately been derived from the latent energy of space itself. Thus, the energizing source for the universe is present everywhere throughout all cosmic space, from which everything that has become visible is derived, and also the plasma particles that are not visible.

## That's the mark of anti-entropy.

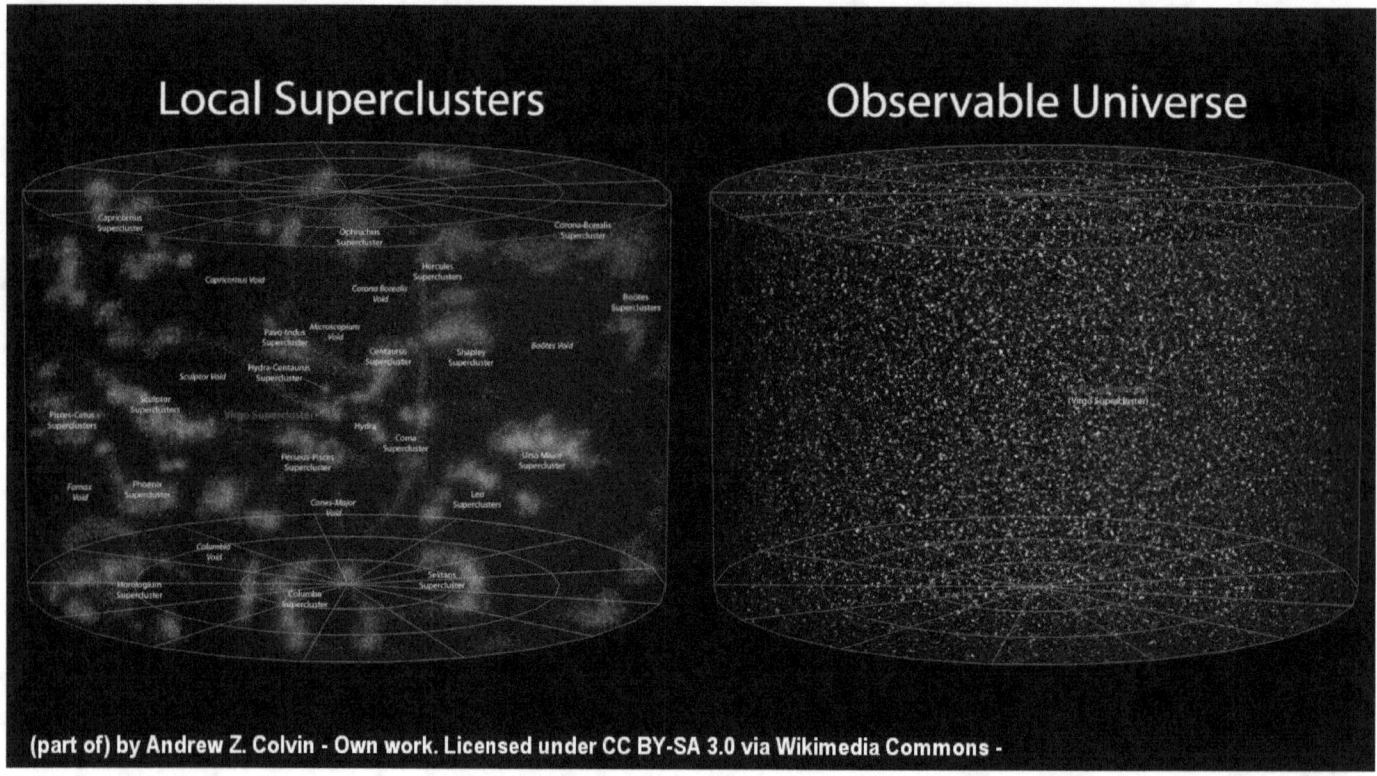

(part of) by Andrew Z. Colvin - Own work. Licensed under CC BY-SA 3.0 via Wikimedia Commons -

With the electromagnetic principles that set the resulting plasma into motion, stars and galaxies of stars are formed everywhere in space, which, by means of plasma fusion, synthesize all the atomic elements that exist in the universe in a creative process that happens everywhere and never ends, by which the universe grows and becomes more massive. That's the mark of anti-entropy.

# Every sun a creator of its surrounding worlds

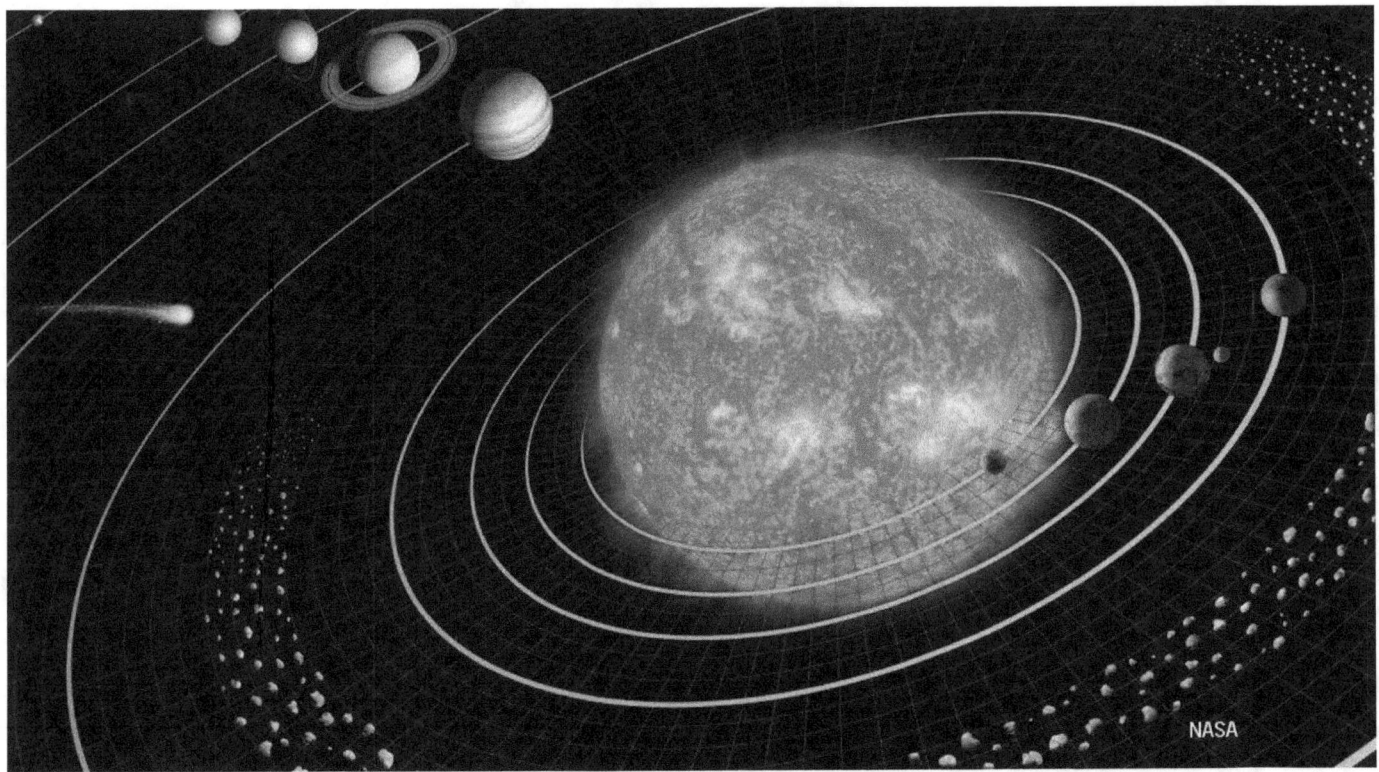

Every sun becomes thereby a creator of its own surrounding worlds. Everything that we see, every atom in the universe, has been created by a sun. This means that a sun that is not an entropic hydrogen-fusion furnace that consumes itself as the Big Bang theory would have it to be, such as a sun condensed from primordial gases - but is instead a plasma sun that is itself the creator of all the hydrogen atoms and other atomic elements that surround the sun and form the solar system. Our Sun, according to all evidence, is a plasma sun and is powered by electric plasma interaction.

# The so-called 'Pillars of Creation'

The so-called 'Pillars of Creation' of the Eagle nebula, are pillars of atomic dust synthesized by a powerful sun located behind each one of the pillars near the top. The Big Bang theory states that the dust of the pillars have created their respective sun as a product of accretion. In the real universe the opposite is true. The evidence is rather simple, that it was the sun behind each pillar that has created the atomic elements that are visible here as clouds of dust and gases. The so-called Pillars of Creation, are pillars of smoke and dust that have been created by an extremely active sun. The sun, thereby, becomes is creative source, rather than the created object, as the Big Bang theory would have it to be.

# Every sun fuses plasma into atoms

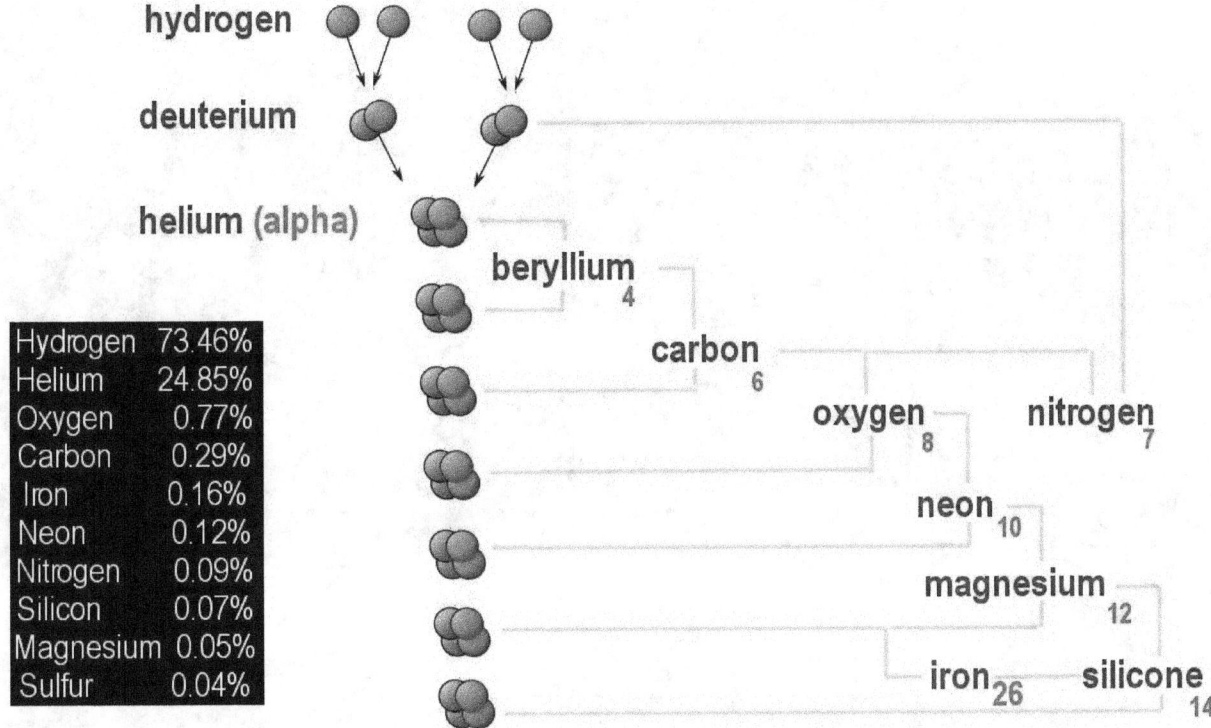

Every sun fuses plasma into atoms. It synthesizes every atomic element that exists. No Big Bang explosion is needed for a universe to be blessed with an abundance of elements. The real universe is self-creating, self-powering, and is self-maintaining and self-advancing, by the dynamics of its timeless principles that have no beginning and end.

## Anti-Entropic Energy

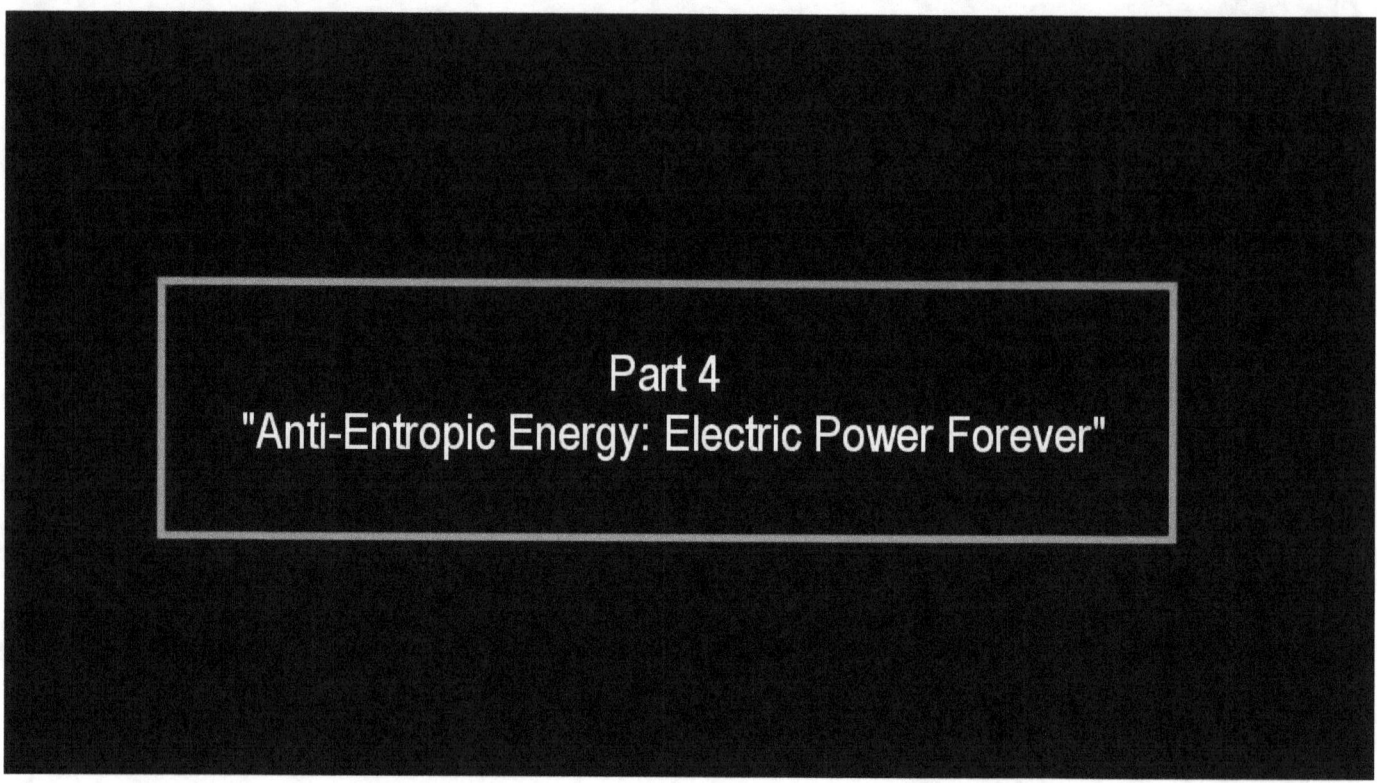

"Anti-Entropic Energy: Electric Power Forever"

# Celebrate that the Big Bang theory is false

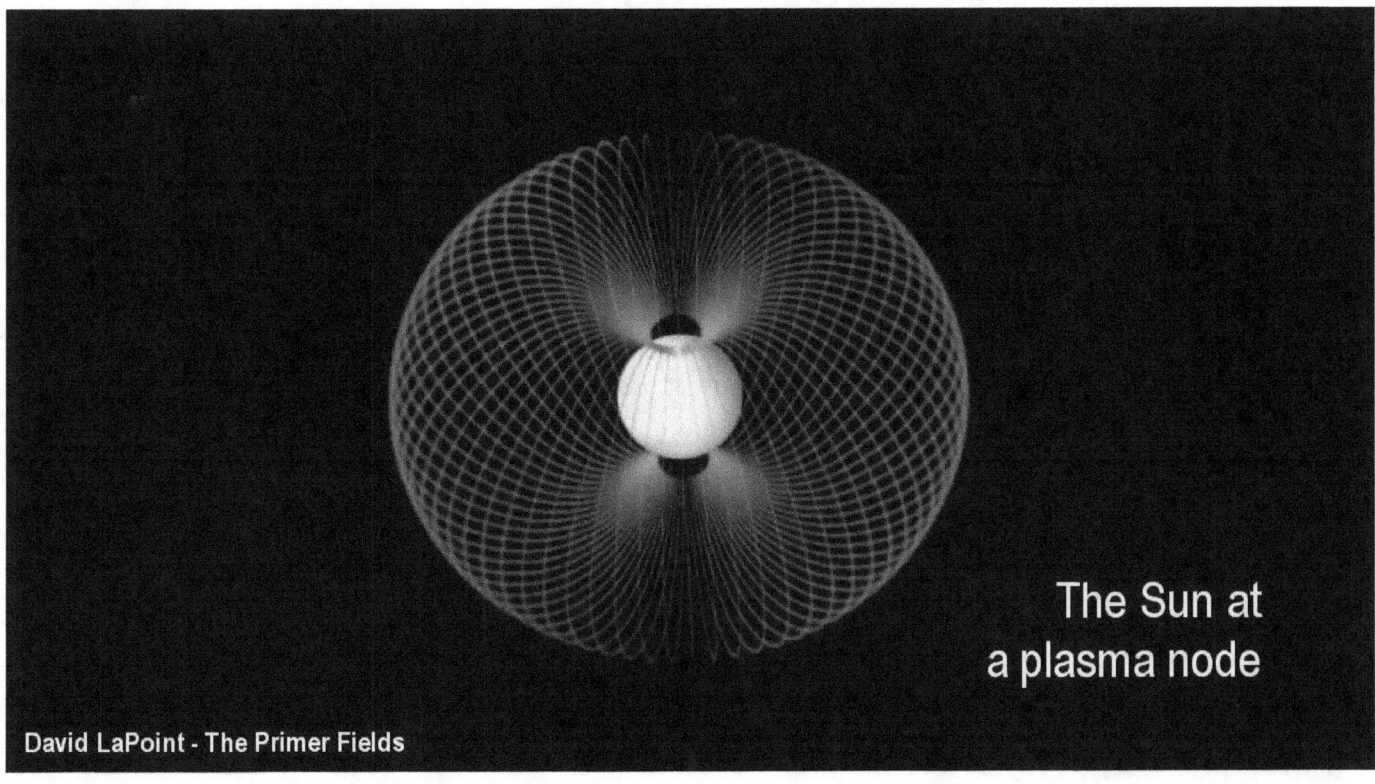

We should celebrate that the Big Bang theory is false. The celebration frees us to celebrate the universe as it is. It enables the recognition that the universe is powered by immensely large, though invisible, cosmic streams of plasma that power every sun, which we too, can access for utilization on the Earth.

# Not a single sun in the universe is self-powered

Not a single sun in the universe is self-powered. Every sun is powered by the universe directly.

# The universe is powered on the cosmic scale

The universe is powered by means of electromagnetic principles on the cosmic scale, that focus electric cosmic plasma streams into a sun.

## Our future depends on this utilization

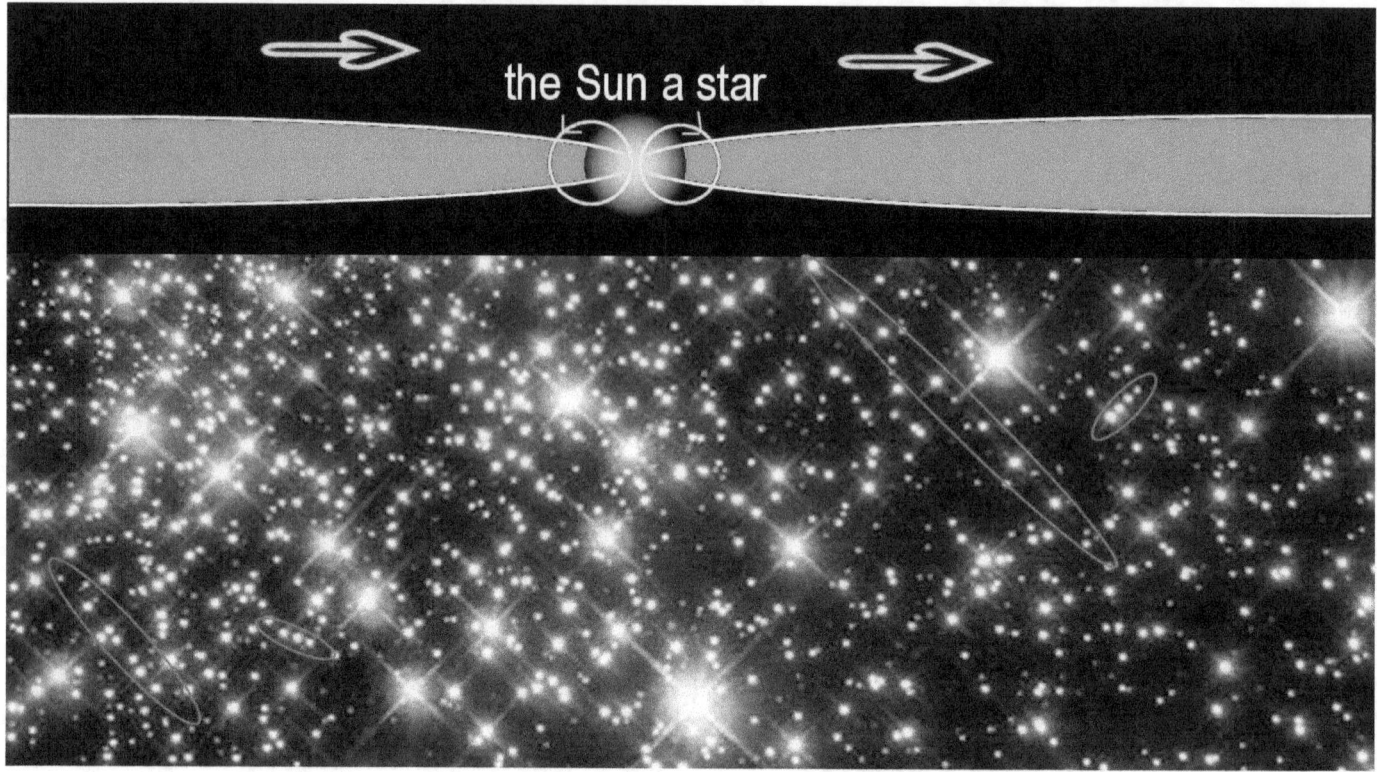

We should celebrate this truth, because the galactic electric energy streams present to us an anti-entropic electric energy resource to power our world with, as we make it available to us. Our future depends absolutely, on this utilization.

# Burning fuels for energy production

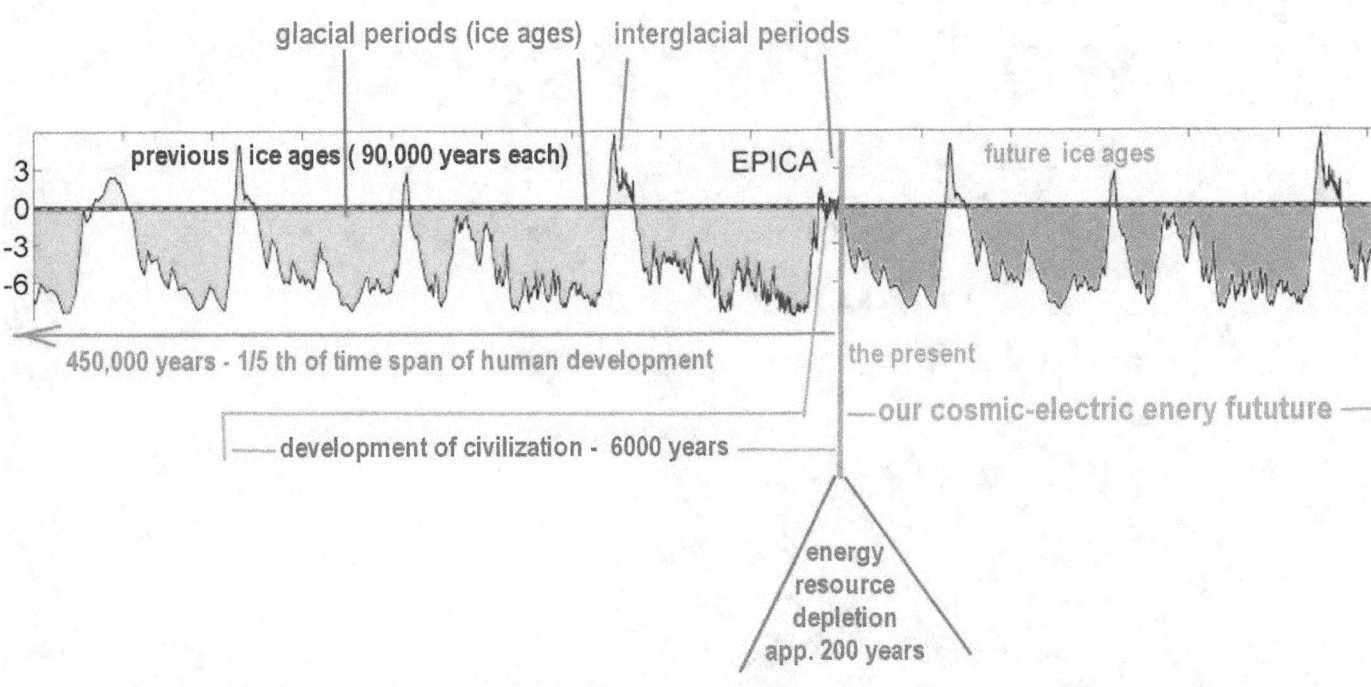

We presently live by burning fuels for energy production. These fuels are being depleted. Oil and gas and nuclear fission fuel may be depleted in 60 years. Coal may last a bid longer. Most of the depletion has occurred in the last 200 years in the 6000 years history of our civilization. Before large-scale energy development began, the only energy resource we had, was wood taken from the land, or oil from fish. The poverty that corresponds with this primitive living, has kept humanity small.

However, once we freed ourselves from the primitive poverty, we found that the needed energy resources that give us our advanced freedom from poverty, are inherently finite. Thus, the question needs to be asked, "what will happen to us when the resources that we rely on have been used up. We may be at this stage in 100 years, or the final depletion may be delayed a 1000 years if we curtail energy use with murderous consequences. When we get to this point, will we lay ourselves down to die? Even if we would restart the nuclear fast-breeder technology that enables the fuller utilization of the available nuclear fuel, by a factor of 20, we would still run out of fuel in roughly 1000 years?

The bottom line is, that our terrestrial fuel resources are so minuscule and finite that we won't have anything left of them 1000 years from now, or potentially much sooner than that.

What will power our economy then, through the next 90,000 years of the coming glaciation period, which will be upon us, potentially, in the 2050s? What will power our world when the presently used energy fuels are depleted? What will power our world for the many millions of years into the future that humanity has the potential to have on the Earth?

## To tap into the cosmic electric energy streams

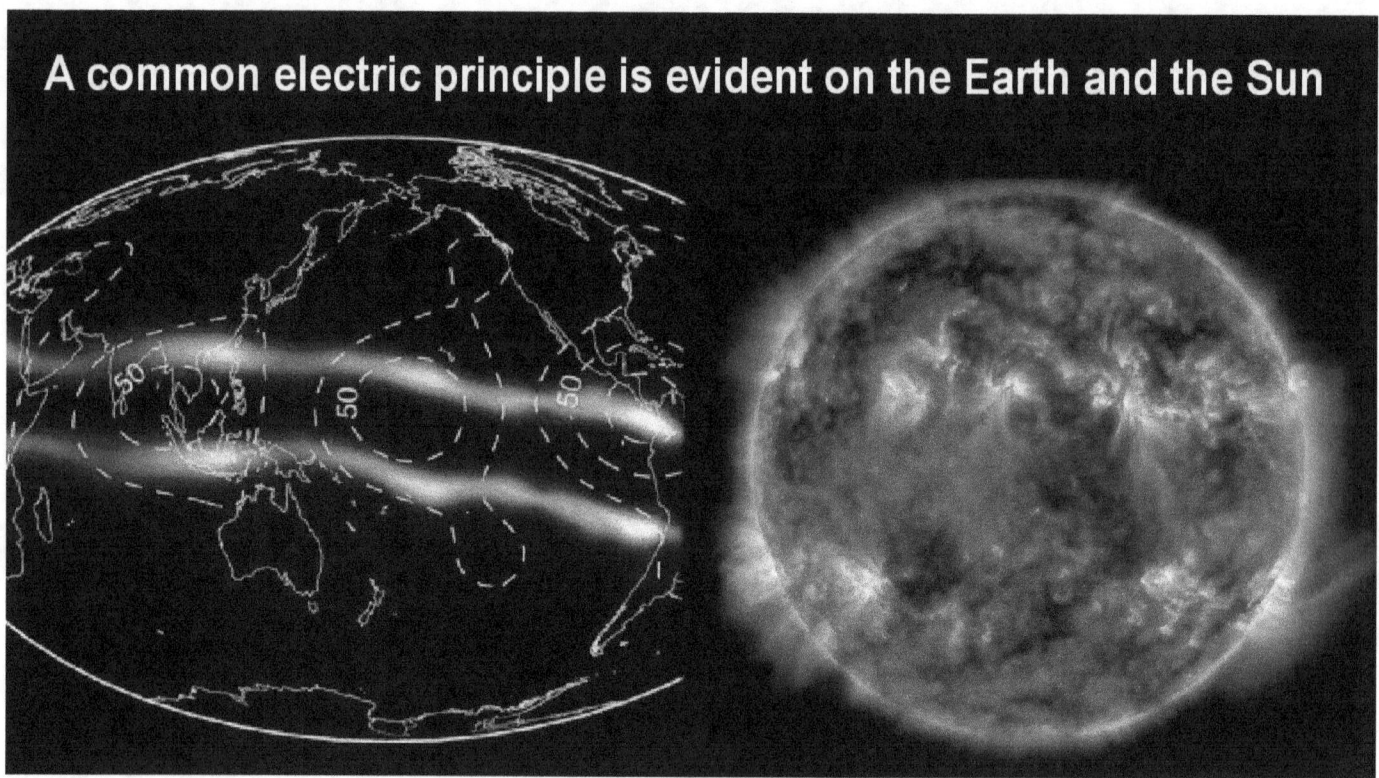

If we didn't have the option to tap into the cosmic electric energy streams that power our Sun, which are evidently available also on the Earth, our future would end at the time we run out of fuels. We would simply die back to minuscule numbers or become extinct.

Fortunately this tragedy does not need to be our future, because the Big Bang theory that blocks our vision of the real universe, is not true.

The cosmic electric energy streams that power every sun, including our Sun, do exist. NASA has even photographed them surrounding the Earth.

## Plasma streams that are encircling the Earth

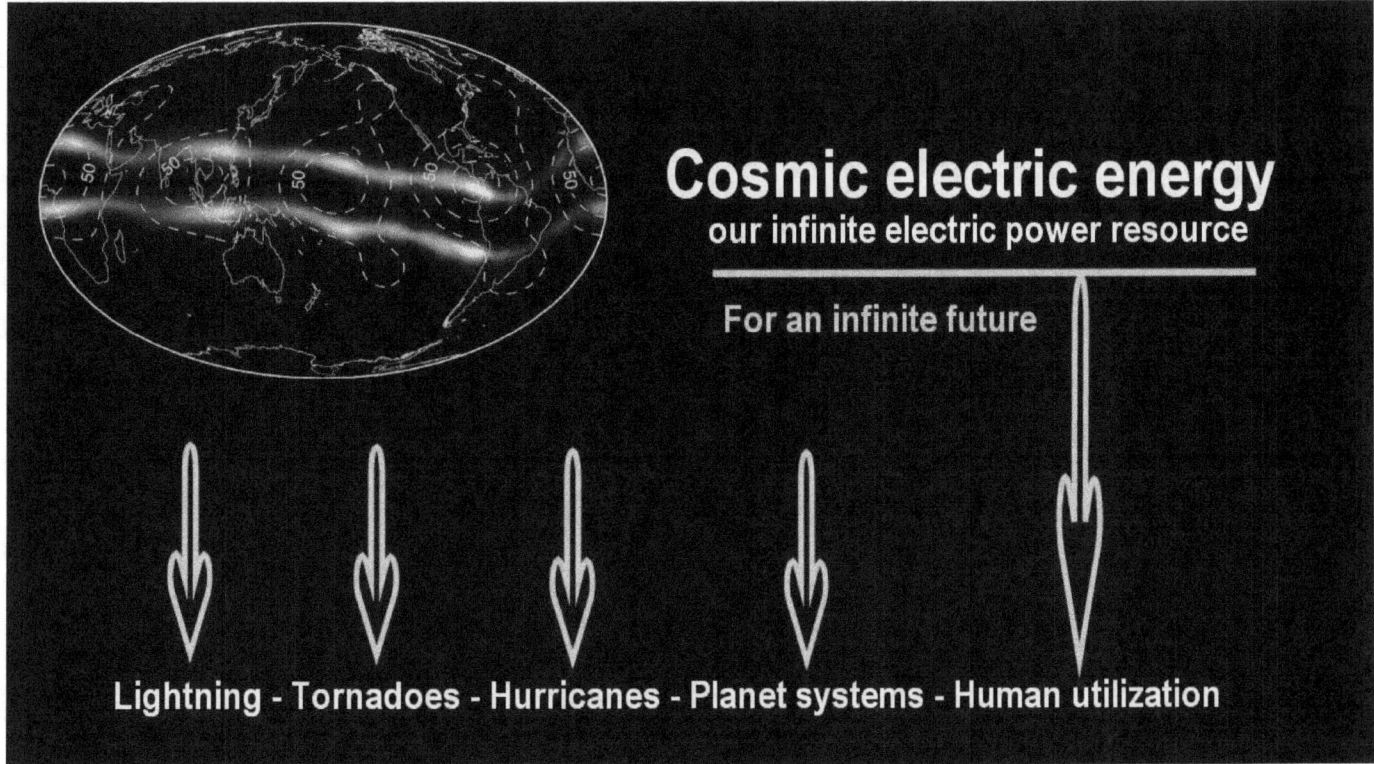

The cosmic plasma streams that are encircling the Earth in the form of two bands centered on the magnetic equator, are cosmic energy streams that promise to be available to us for our future. Of course, the technology for the utilization of this resource needs yet to be developed. And this is only a technological step away. The Big Bang theory, in contrast, would have us believe that a cosmic energy resource outside the bounds of the Earth doesn't exist. Fortunately, the theory is wrong. It has no foundation. Let's celebrate that the truth is much grander, which gives us an infinite future.

## The Big Bang Without a Future

"The Big Bang Without a Future"

# The Big Bang theory insists

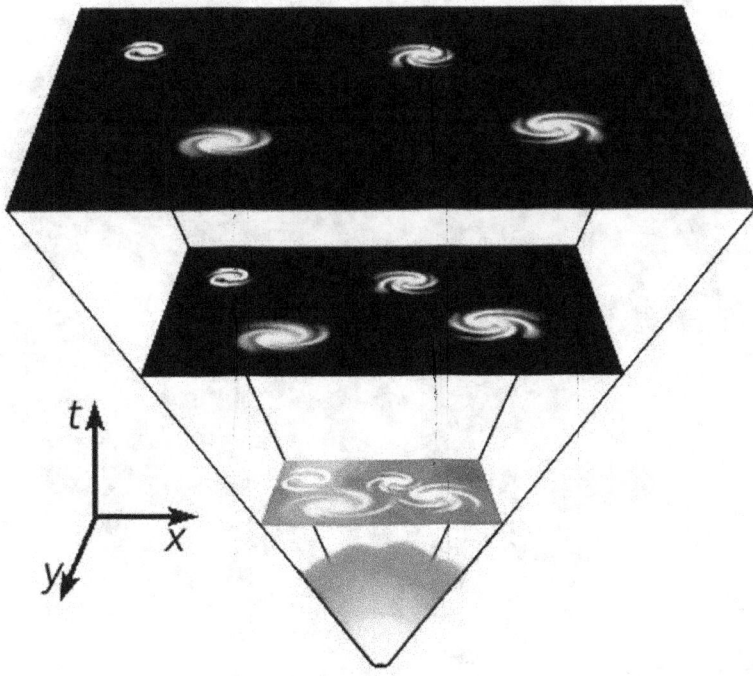

The Big Bang theory insists that humanity has no future; that the universe itself has no future. It insists that the universe was given one single shot of energy, one single infusion, and that's all it got, which implies that our tiny portion on the Earth is the totality that we will ever have.

# When the energy resources are gone, our future ends

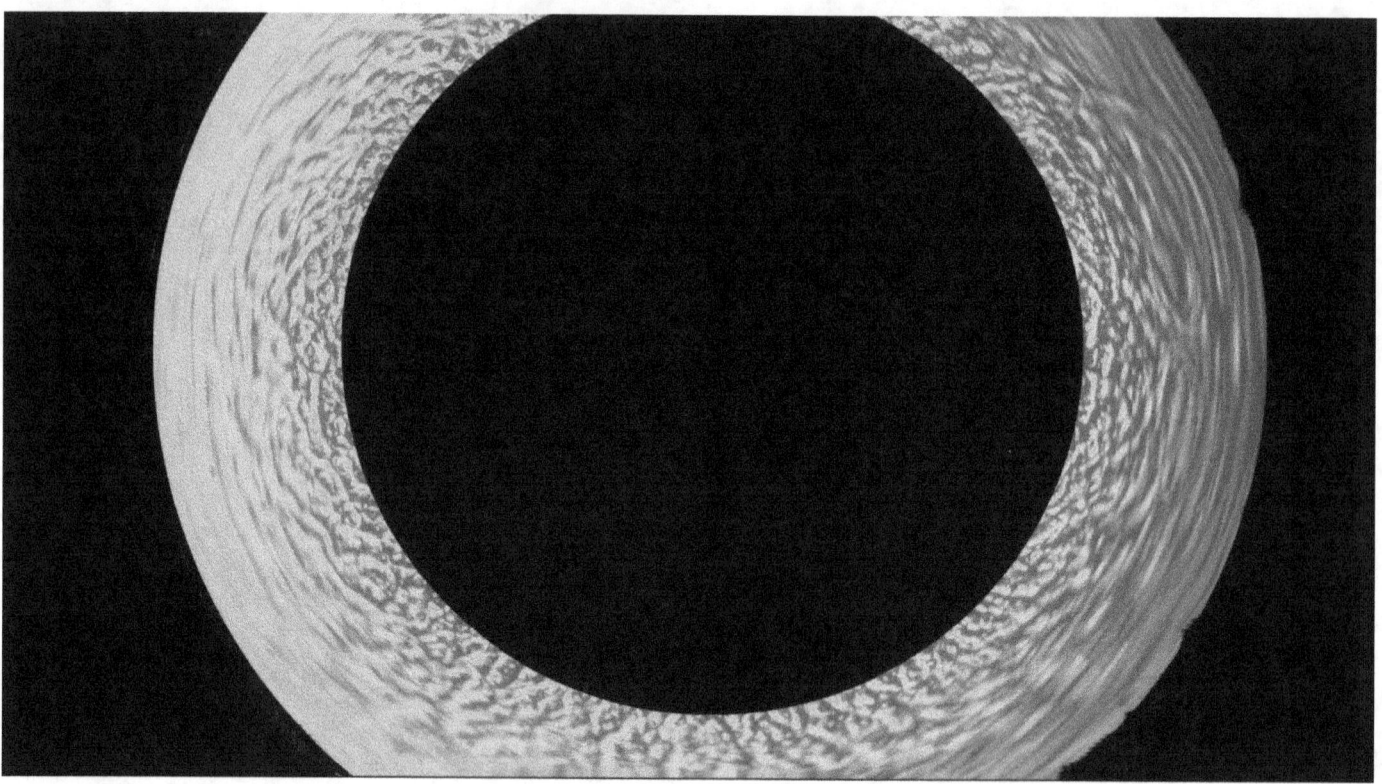

The theory says that energy was spread across the universe by the big explosion, so that what we got from it, is all we'll ever have, so that by our burning it, our future is diminishing. In other words: when the energy resources are gone, our future ends.

# The concept by which everything ends

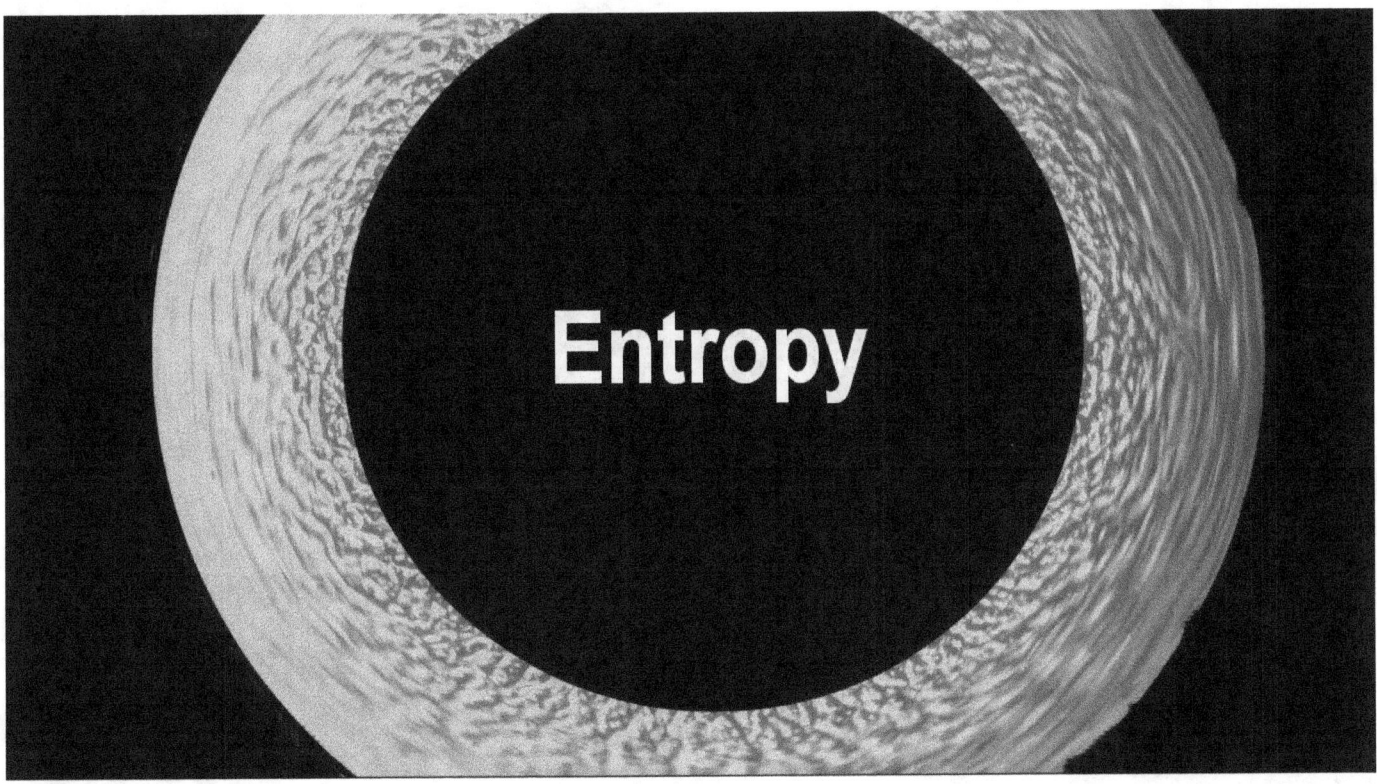

The concept by which everything ends, is termed, entropy. The Big Bang theory is entropic, because it says, that once we have used up what we were given, we will have nothing and we will die.

## Even the Sun will die

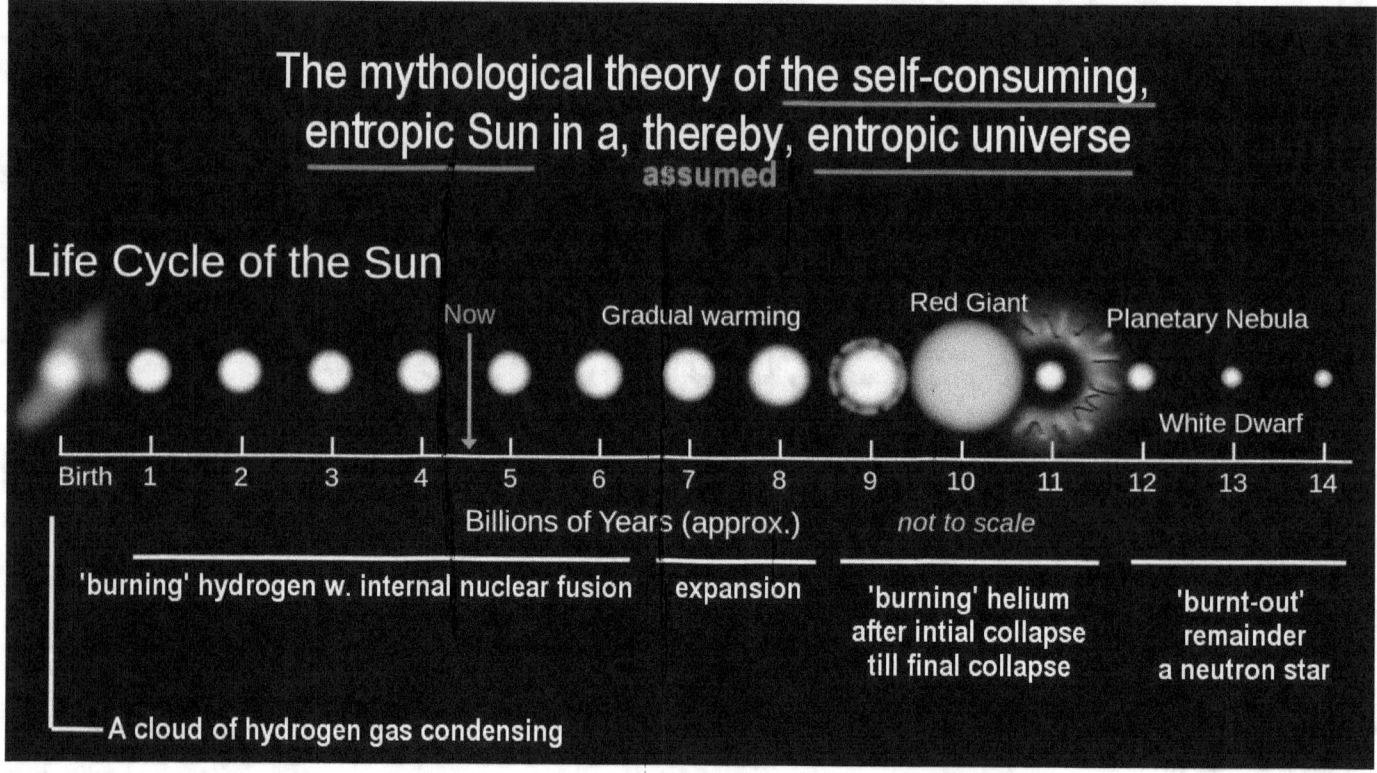

The Big Bang theory says that even the Sun will die when it used up its fuel. It says that the entire universe will grind down to nothing and die the inevitable energy-depletion death.

# Fortunately, we see no evidence

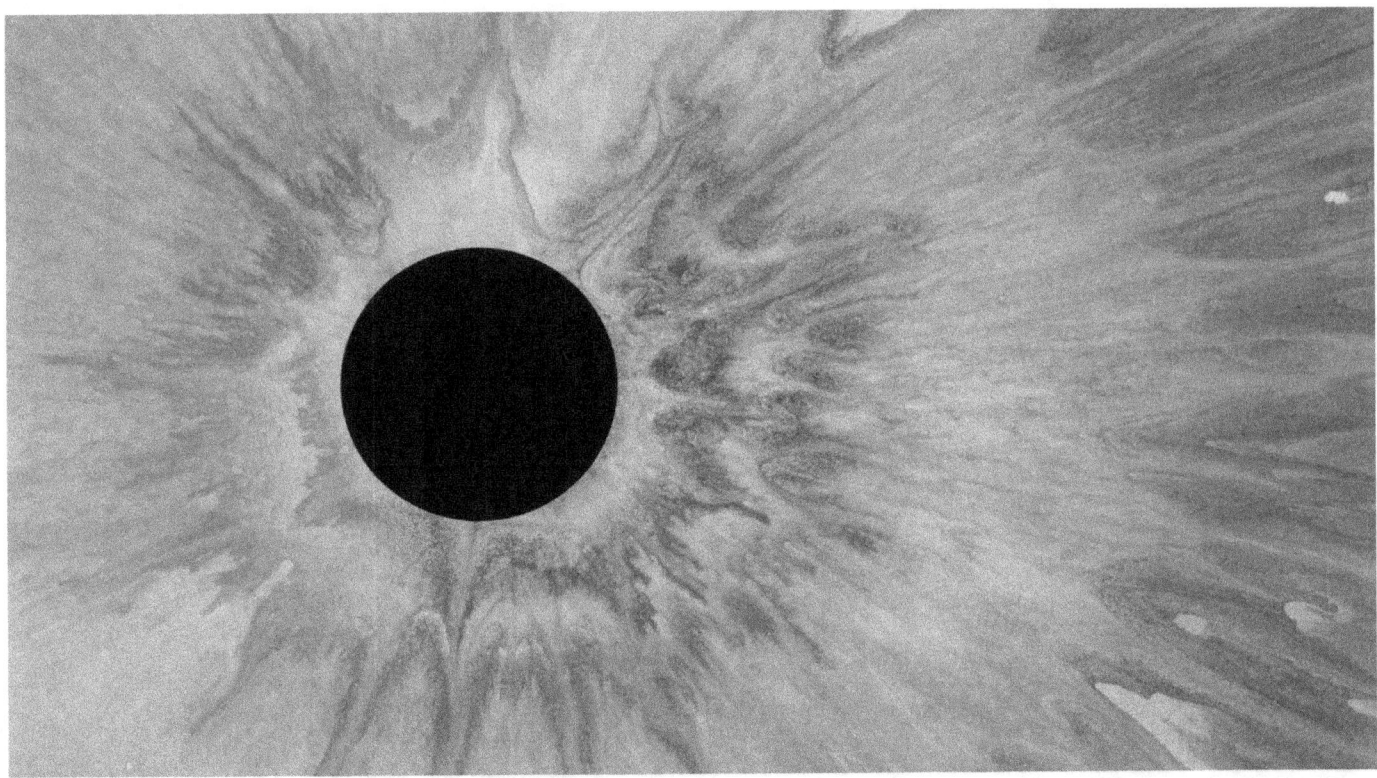

Fortunately, we see no evidence that the theory is even remotely credible. If a big explosion had created all matter in the universe, which the explosion expanded and forged with it all visible forms, then the distribution of matter would have become diffused with the cube of the distance, according to the dynamics of an explosion. This means that the primordial matter would have become spread so thin farther out that almost nothing would be found in distant places, much less the gigantic volumes of it that supposedly condense into clusters of galaxies. If the Big Bang theory was correct, there souldn't be anything much to be seen in the universe. Of course, this is not how the universe reveals itself to the viewers with telescopes.

## The observable universe is uniformly dense

The center of the Milky Way, at the center of the Big Bang explosion of the universe

The observable universe is uniformly dense with near even distribution of galaxy super-clusters throughout the vast reaches of space that telescopes can observe, ranging from the local super clusters to the vast sea of super clusters that are spread out across the infinite realms of the cosmos.

# We are seeing a universe that is actively self-powered

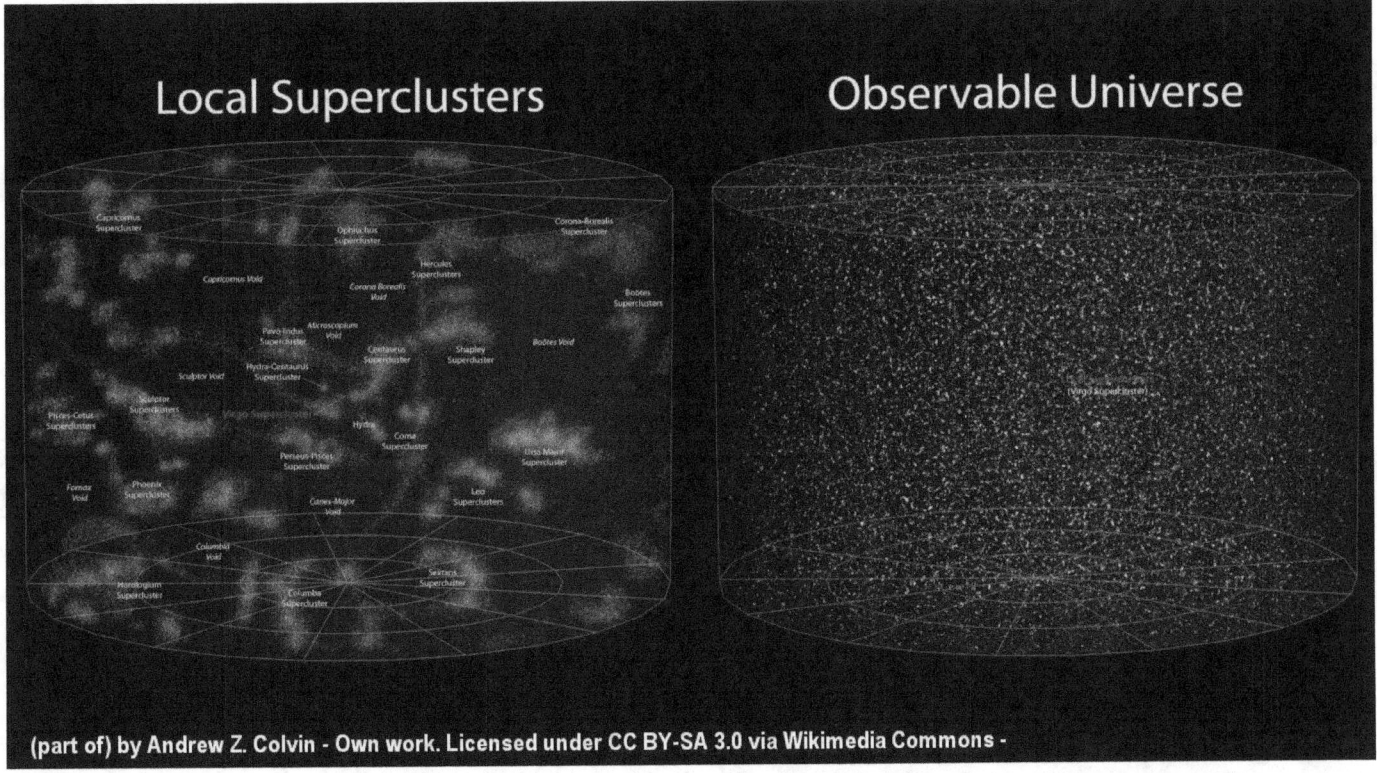

We are looking at a universe with our telescopes, that is evidently self-created everywhere, instead of having originated from a single-point source. We are seeing a universe that is actively self-powered, and is actively self-maintained, and this everywhere with nothing running down.

But how is this possible? How can an entire universe be self-powered?

# The Anti-Entropic, Self-Powered Universe

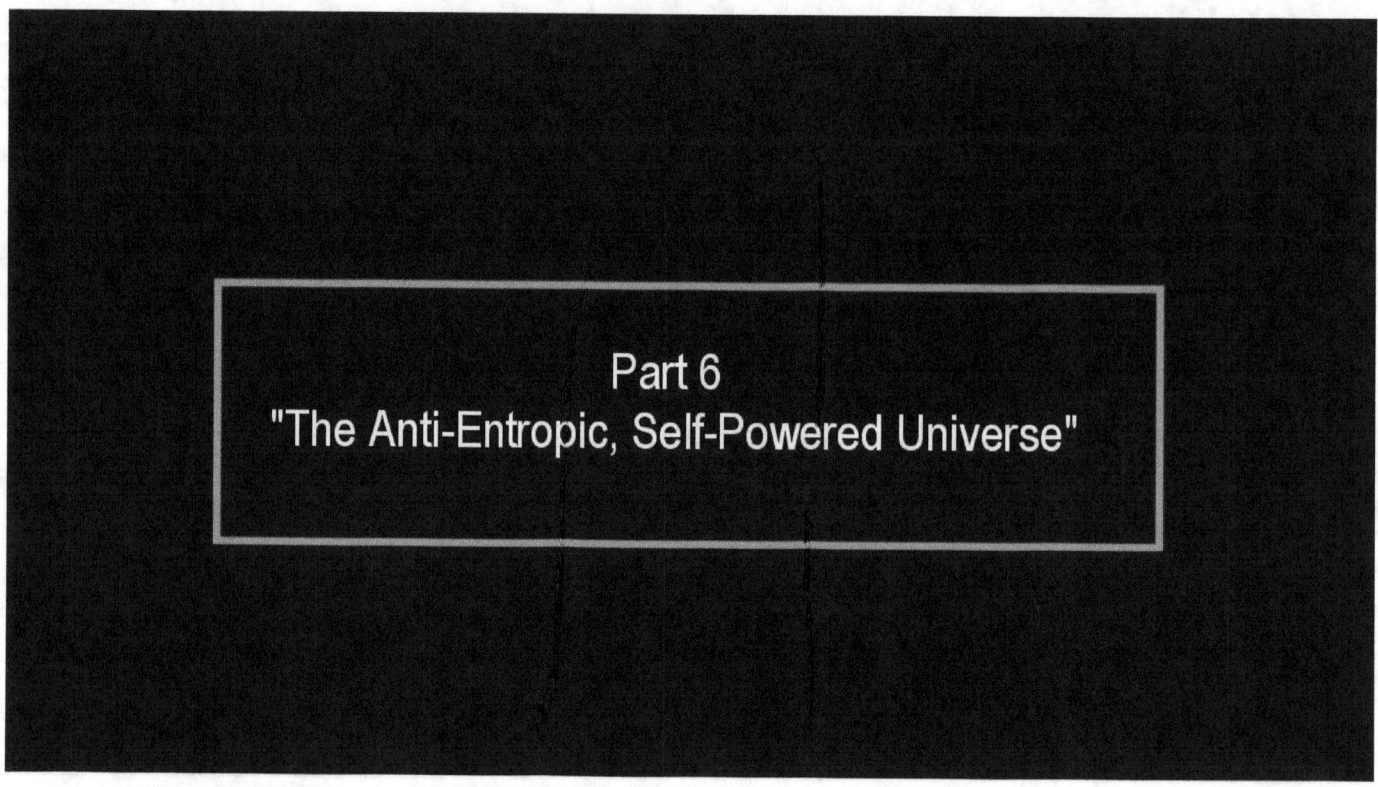

"The Anti-Entropic, Self-Powered Universe"

# Back to David Bohm

This subject takes us back to David Bohm, whom Einstein is said to have called his successor. As I said earlier, David Bohm speaks of cosmic space as not being empty, but as being a vast sea of latent energy with an implicate order and an explicate order. But what does this mean?

# A latent sea of water

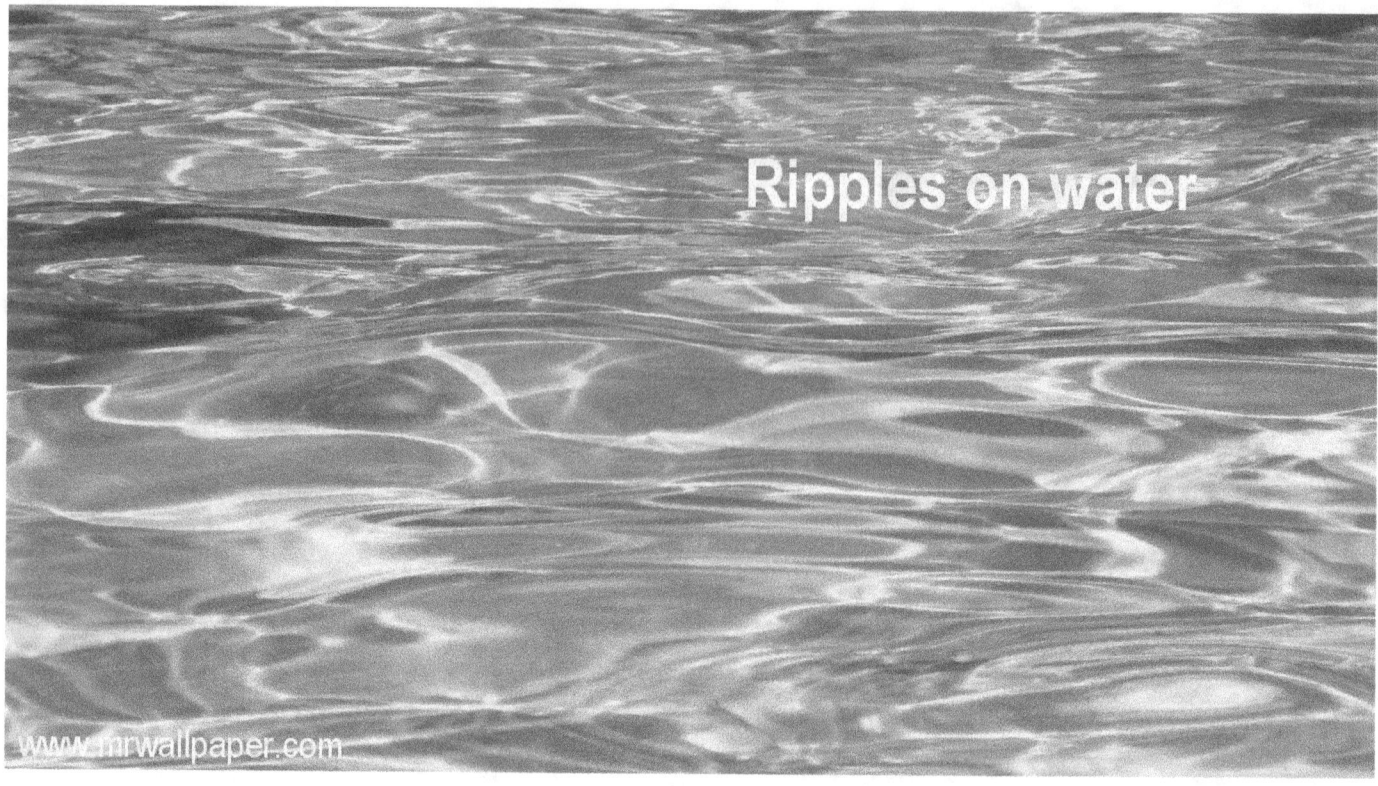

Our oceans can be seen as a latent sea of water. Nothing much happens there. However, a tiny portion of the water constantly evaporates at the surface.

## Water vapor

The water vapor, though, is too small in size to be visible by itself. Only when water molecules latch together and form tiny droplets, will the evaporated vapor become faintly visible as fog.

# The latent background of energy

## Standard Model of Elementary Particles

| | mass | charge | spin | | | |
|---|---|---|---|---|---|---|
| **QUARKS** | ≈2.3 MeV/c² / 2/3 / 1/2 **u** up | ≈1.275 GeV/c² / 2/3 / 1/2 **c** charm | ≈173.07 GeV/c² / 2/3 / 1/2 **t** top | 0 / 0 / 1 **g** gluon | ≈126 GeV/c² / 0 / 0 **H** Higgs boson | |
| | ≈4.8 MeV/c² / -1/3 / 1/2 **d** down | ≈95 MeV/c² / -1/3 / 1/2 **s** strange | ≈4.18 GeV/c² / -1/3 / 1/2 **b** bottom | 0 / 0 / 1 **γ** photon | | |
| **LEPTONS** | 0.511 MeV/c² / -1 / 1/2 **e** electron | 105.7 MeV/c² / -1 / 1/2 **μ** muon | 1.777 GeV/c² / -1 / 1/2 **τ** tau | 91.2 GeV/c² / 0 / 1 **Z** Z boson | | **GAUGE BOSONS** |
| | <2.2 eV/c² / 0 / 1/2 **νe** electron neutrino | <0.17 MeV/c² / 0 / 1/2 **νμ** muon neutrino | <15.5 MeV/c² / 0 / 1/2 **ντ** tau neutrino | 80.4 GeV/c² / ±1 / 1 **W** W boson | | |

by MissMJ - Own work by uploader, PBS NOVA, Fermilab, Office of Science, U.S. Department of Energy, Particle Data Group. Licensed under CC BY 3.0 via Wikimedia Commons

The latent background of energy that David Bohm speaks about, may be likened in comparison to a vast ocean of energy, from which tiny parts become discrete, become explicate, like molecules of water vapor that become discrete from the oceans. In space, the discrete explicates of the latent energy become the basic elementary particles of the universe. The major groups of these explicate discrete entities of energy, are termed quarks and leptons. The quarks have built-in characteristics that combines into specific forms, that form the proton particle, and in a similar manner the leptons for the the electron particle.

# The electron

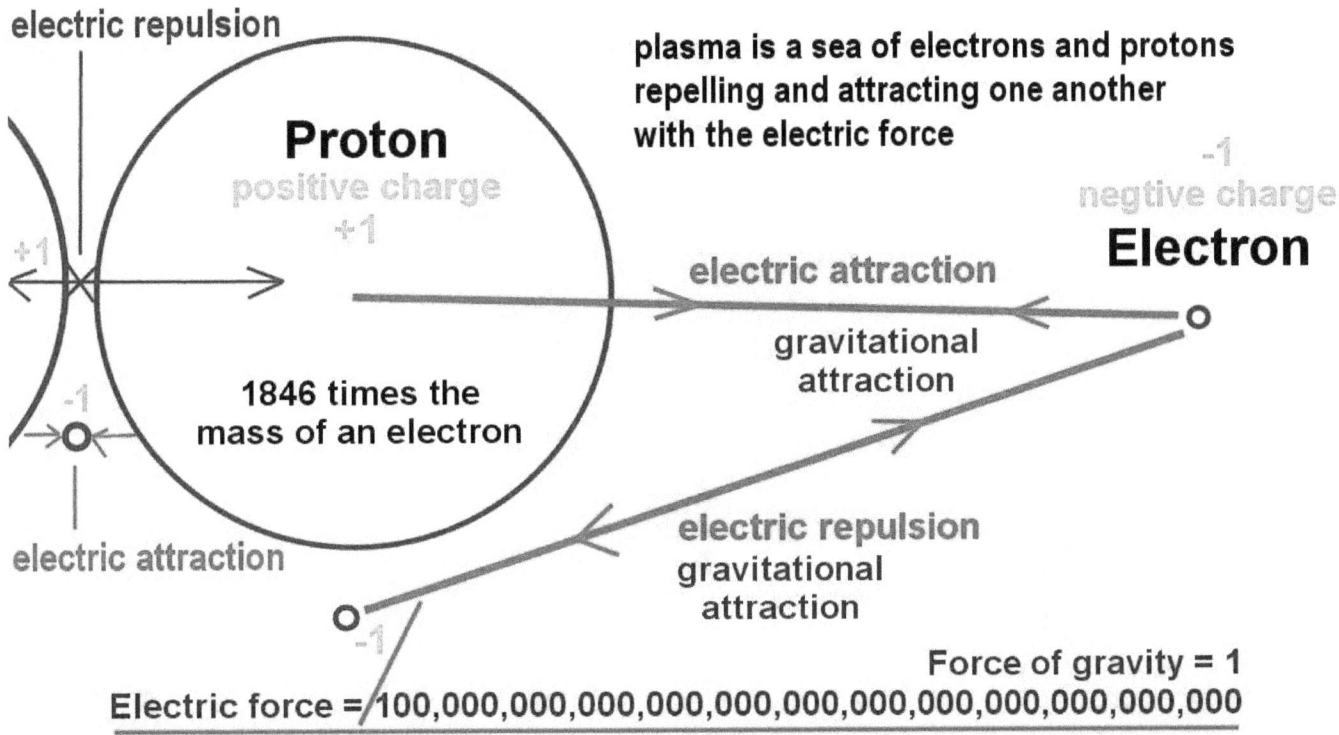

The electron is the smallest of the particles. It is a thousand times smaller than a proton.

# A proton is made up of three quarks

## The quark structure of the proton

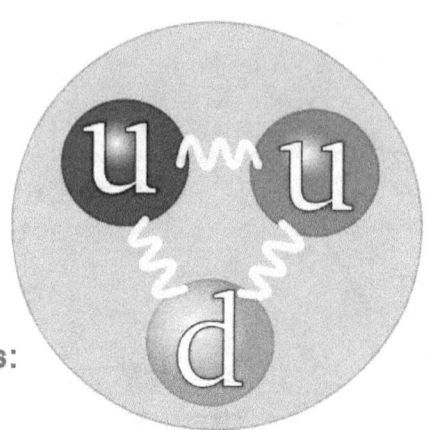

The proton is made up of 3 quarks:
2 up quarks (2/3 charge)
1 down quark (-1/3 charge)
Held together by the strong force
(its total charge = 1)
Its mass is dynamic (80 to 100 times the rest mass of the quarks
The strong force is mediated by gluons (wavey)

"Quark structure proton" by Arpad Horvath - Own work. Licensed under CC BY-SA 2.5 via Wikimedia Commons -

A proton is made up of three quarks that are bunched together, which gives the proton a substantial size, and a substantial apparent mass.

## The electron, in comparison

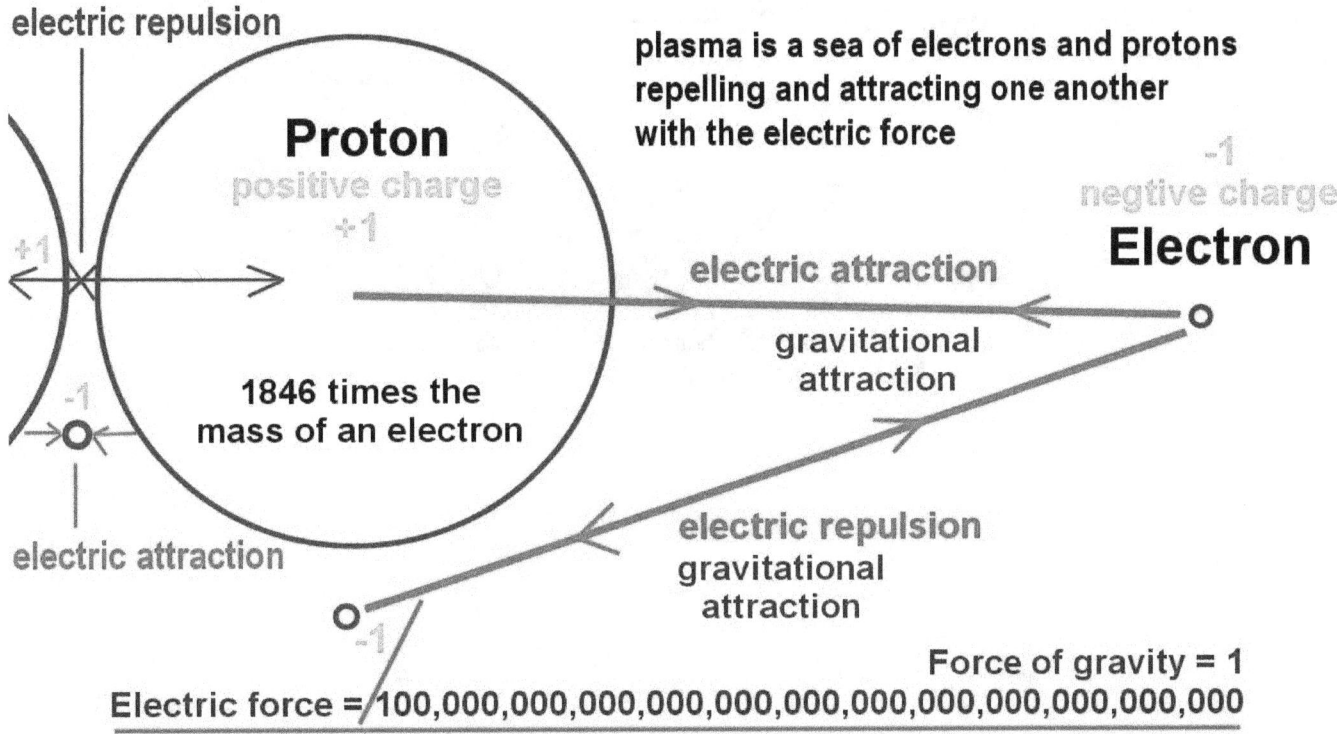

The electron, in comparison, is so small that it exists partly as but an energy wave, and only partly as a particle.

Segment 1- Entropy versus Anti-Entropy

## Quarks and leptons are energy in motion

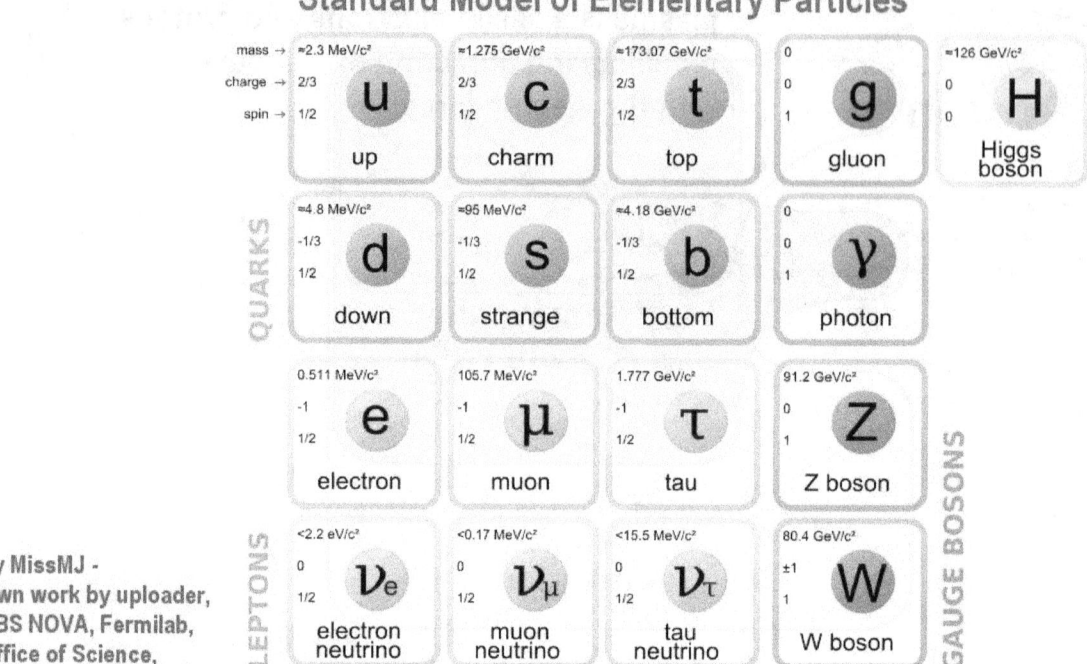

This means that the quarks and leptons, which are but energy in motion, are the basic substance for the basic building blocks of the universe.

## Protons and electrons exist in space

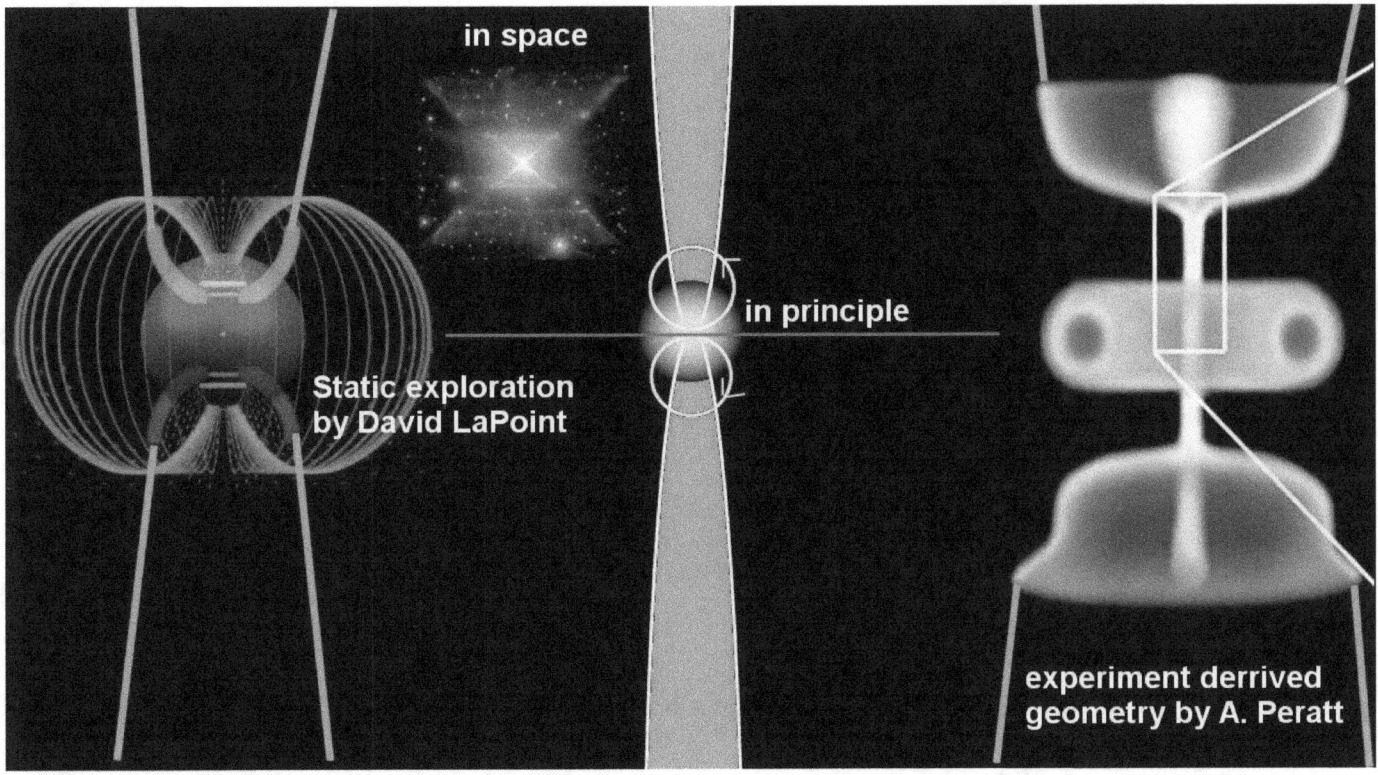

The protons and electrons exist in space, organized into vast streams termed plasma. The plasma in turn carries the forces by which the visible universe is formed and organized, and gains its mass.

## Roughly 99.94% of the mass of the universe

### The quark structure of the proton

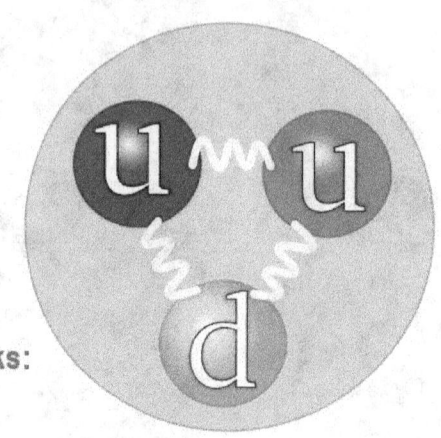

The proton is made up of 3 quarks:
2 up quarks (2/3 charge)
1 down quark (-1/3 charge)
Held together by the strong force
(its total charge = 1)
Its mass is dynamic (80 to 100 times the rest mass of the quarks
The strong force is mediated by gluons (wavey)

"Quark structure proton" by Arpad Horvath - Own work. Licensed under CC BY-SA 2.5 via Wikimedia Commons -

Since roughly 99.94% of the mass of the universe is provided by the protons, and the protons are made of quarks that are theorized entities of energy, it may be useful to look at the protons once more, and the quarks that form them. The quarks clump together in groups of three. Each carries a specific electric charge, which, when they are grouped together, give the resulting proton a positive electric charge, which in turn complements precisely an electrons negative charge. Without this precise balance, which is essential in constructing the visible universe that is made up of atomic elements, the visible universe and its higher-order constructs, with intelligent life at its pinnacle, would not exist.

## Plasma, the lifeblood of the universe

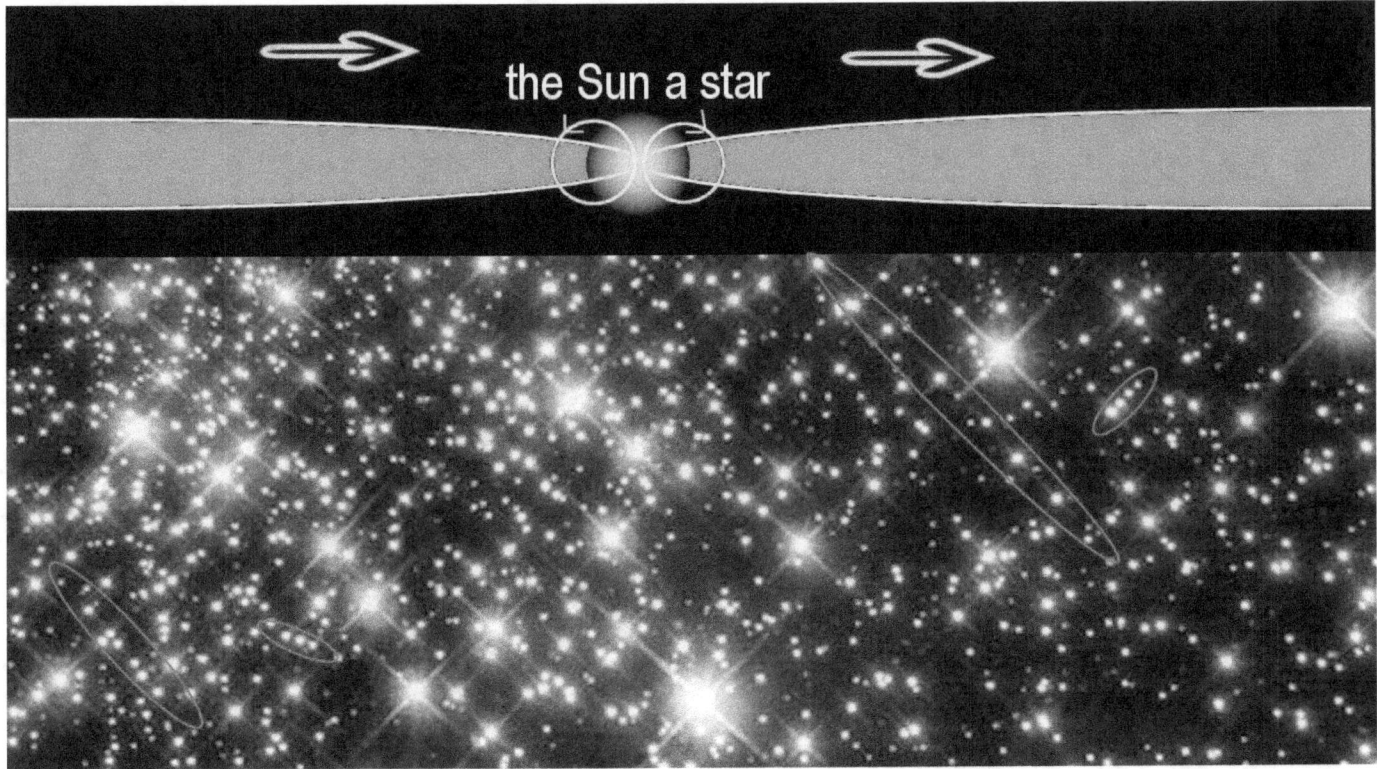

This means that the substance of the universe in all its forms is energy that exists simply everywhere, and is constantly created everywhere. Thus, the pure energy entities that unfold as quarks and leptons, are through the constructs of plasma, the lifeblood of the universe.

## The universe without plasma is inconceivable

This 'blood' of energy flows everywhere, from which everything is created. The universe without plasma is inconceivable, except in dreams. It renders the universe as anti-entropic in nature.

## All space is filled with plasma

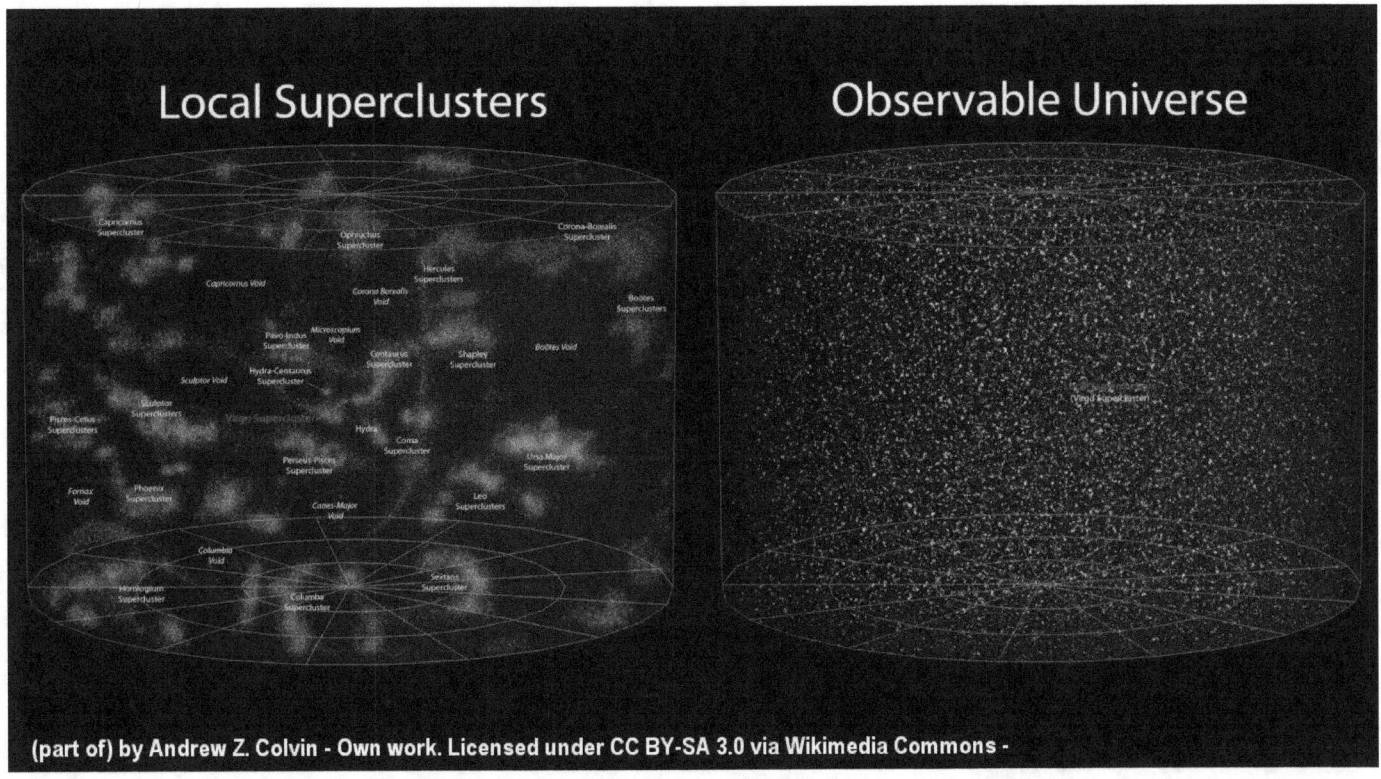

(part of) by Andrew Z. Colvin - Own work. Licensed under CC BY-SA 3.0 via Wikimedia Commons -

All space is filled with plasma. It combines into giant streams. The streams of plasma have formed every sun, and continue to do so. Plasma also carries the energy that powers every sun and by which plasma becomes fused in high energy processes on the surface of a sun in an electric synthesis where the atoms of the universe are forged. This process happens now, and always had, and always will.

Segment 1- Entropy versus Anti-Entropy

# Every atom that exists

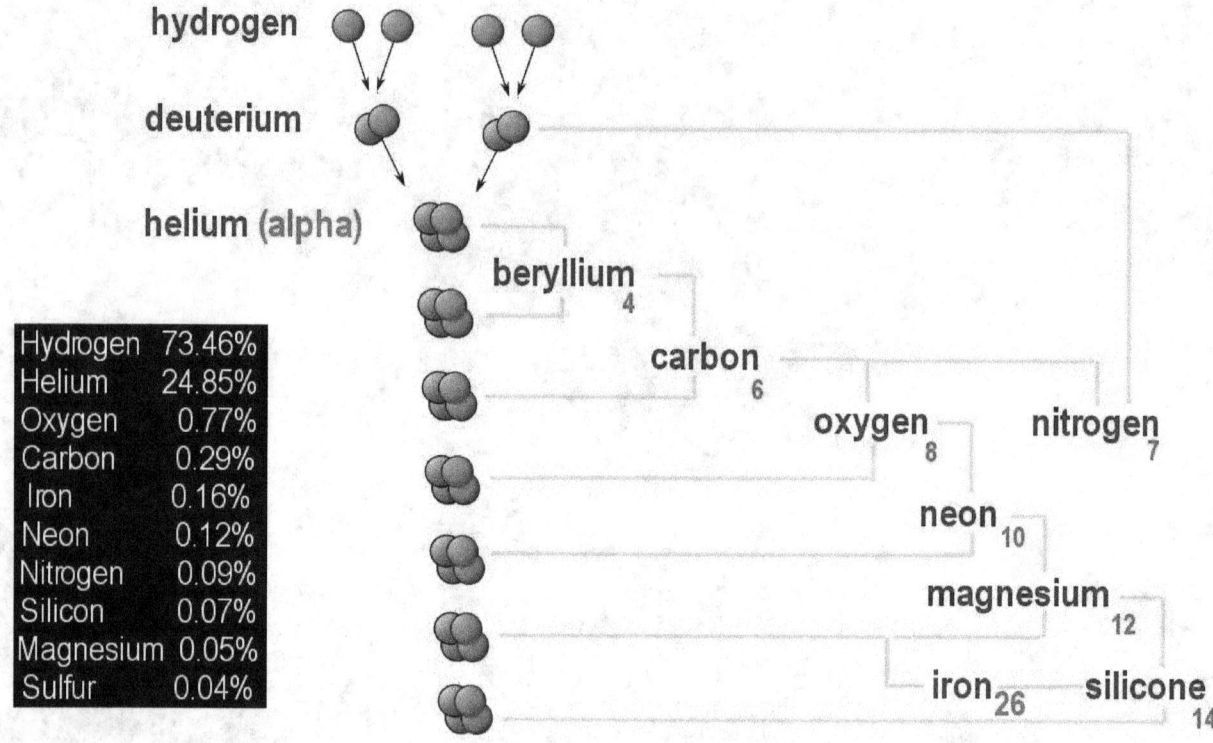

Every atom that exists in the universe was synthesized from plasma in electric fusion on a solar surface.

## The Big Bang creation theory

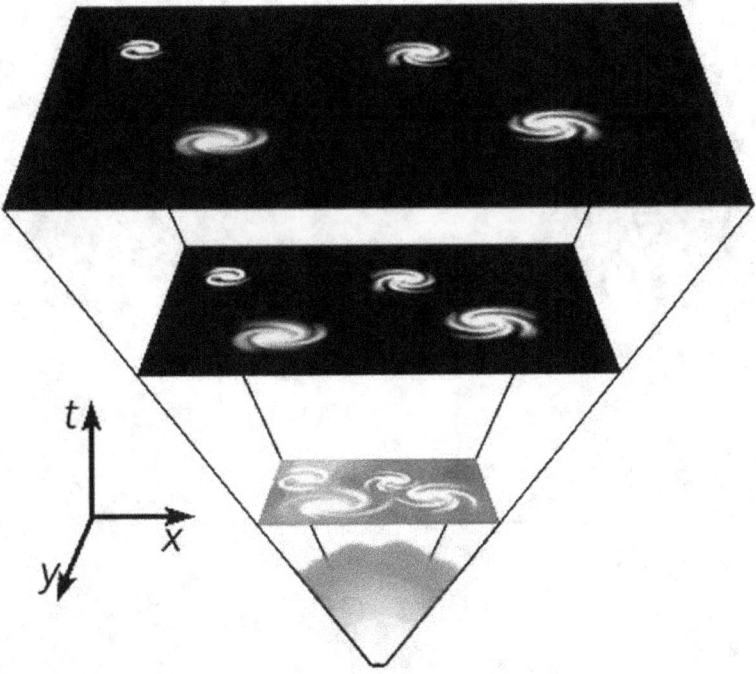

The Big Bang creation theory is nothing more than just a dream.

## The creative process

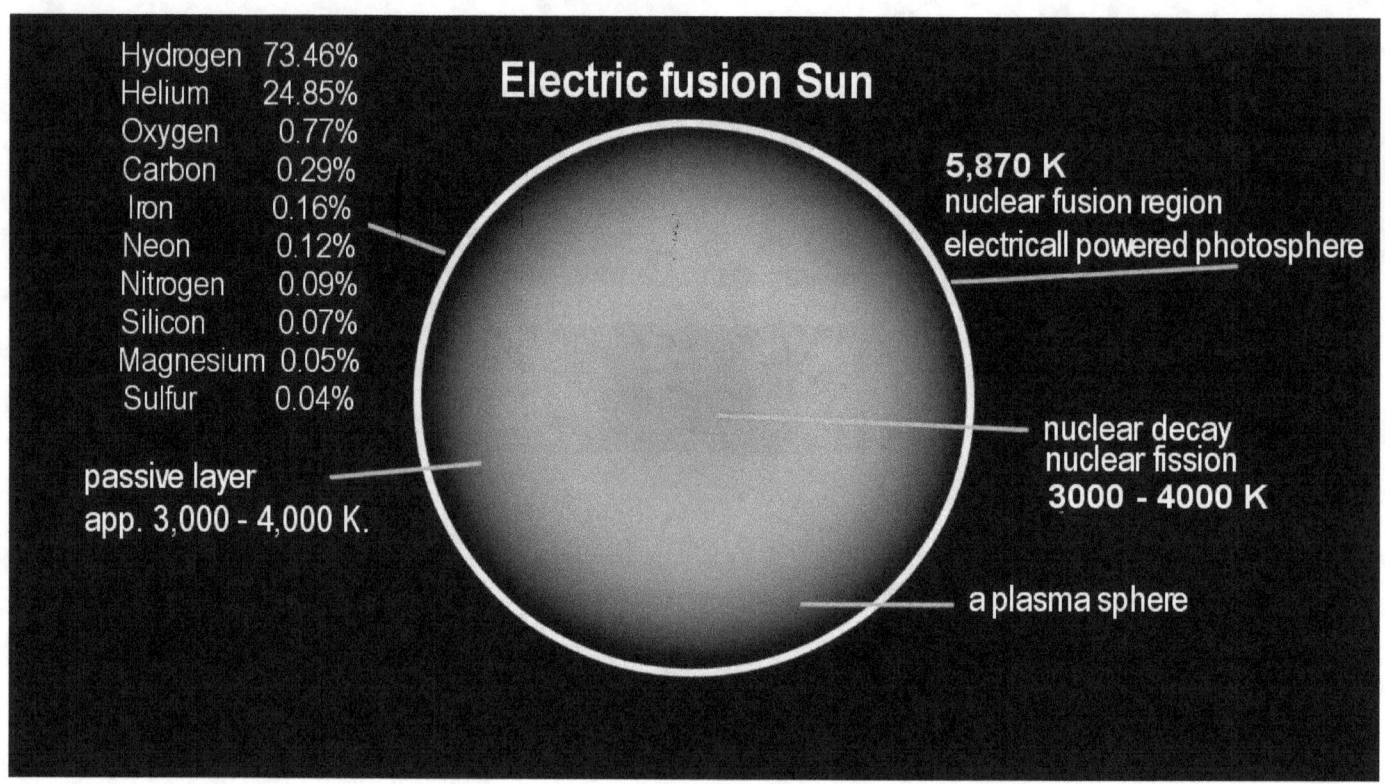

The creative process, of course, continues. It continues everywhere in the universe simultaneously. It is anti-entropic.

# Nothing ever had a single-point origin

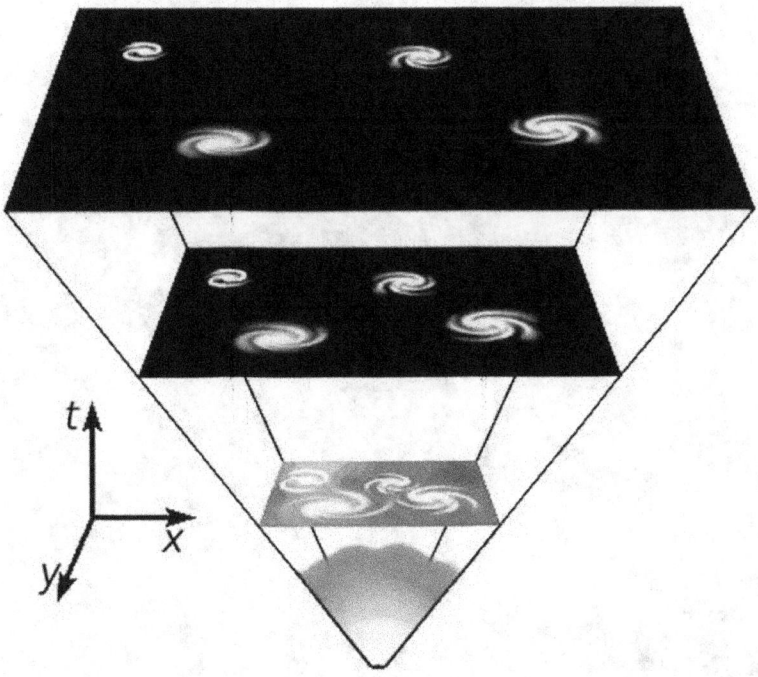

Nothing ever had a single-point origin, and collapse as its destiny.

## The creative process is universal

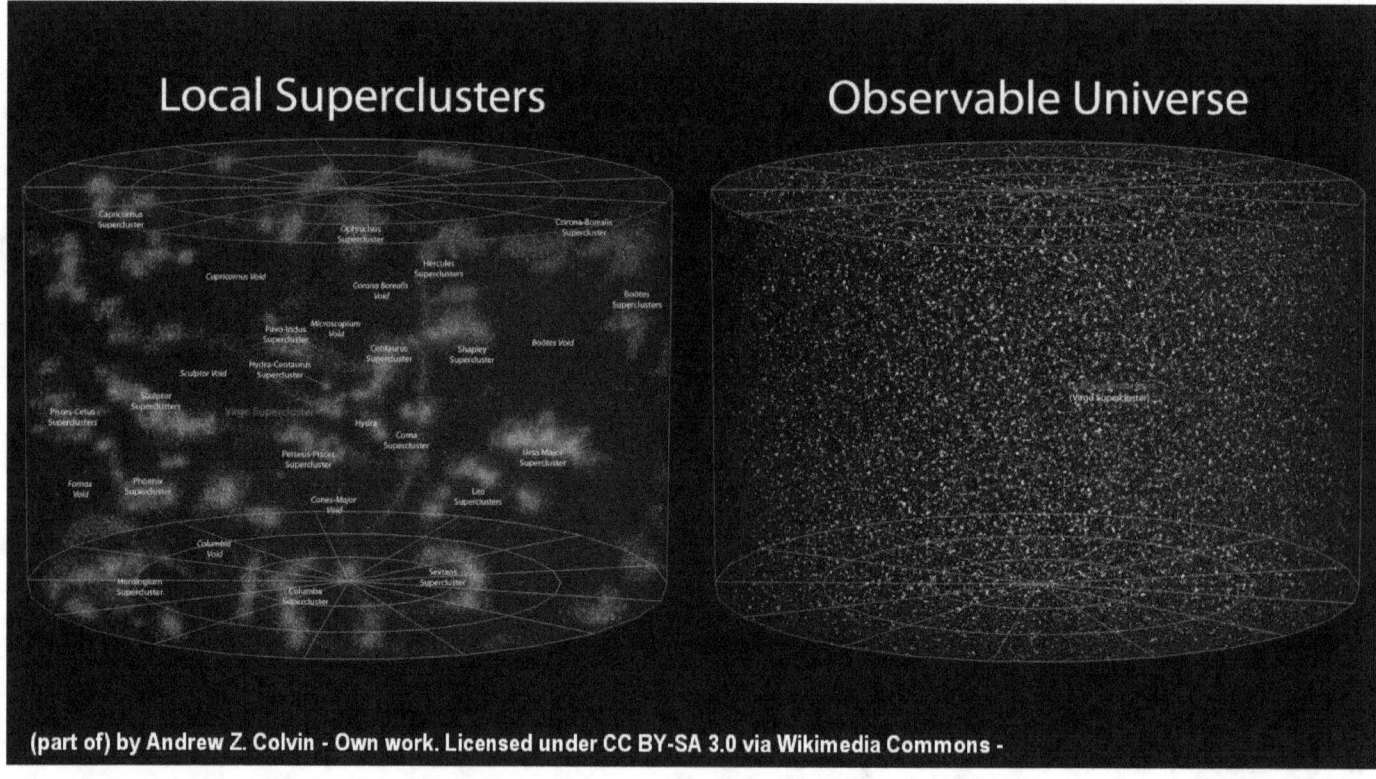

(part of) by Andrew Z. Colvin - Own work. Licensed under CC BY-SA 3.0 via Wikimedia Commons -

The creative process is universal. What you see here is what an anti-entropic universe necessarily looks like, and does look like.

# Nuclear-fusion energy production

All this also means that the fabled nuclear-fusion energy production that is pursued on Earth, is not a viable option. The theorized basis for nuclear-fusion power is wrong, because the Sun, that it aims to replicate, is not inherently an energy producer, but acts as an energy converter of the energy that flows in the universe.

## A sun is not entropic in nature

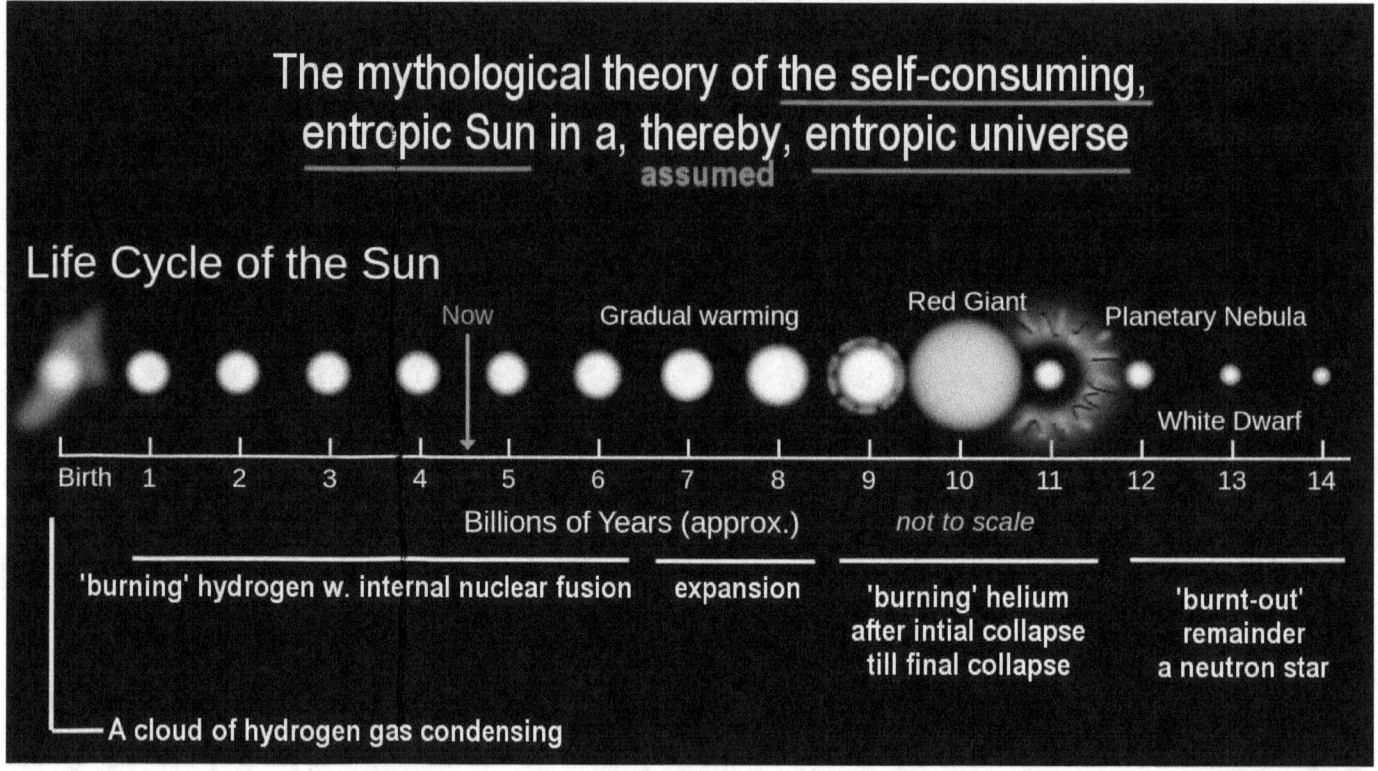

A sun is not entropic in nature. It is not self-consuming to produce energy. An energy-producing sun does not exist. The nuclear- fusion energy production that aims to replicate the energy production of the Sun is a mistake in premise.

# The universe is energy

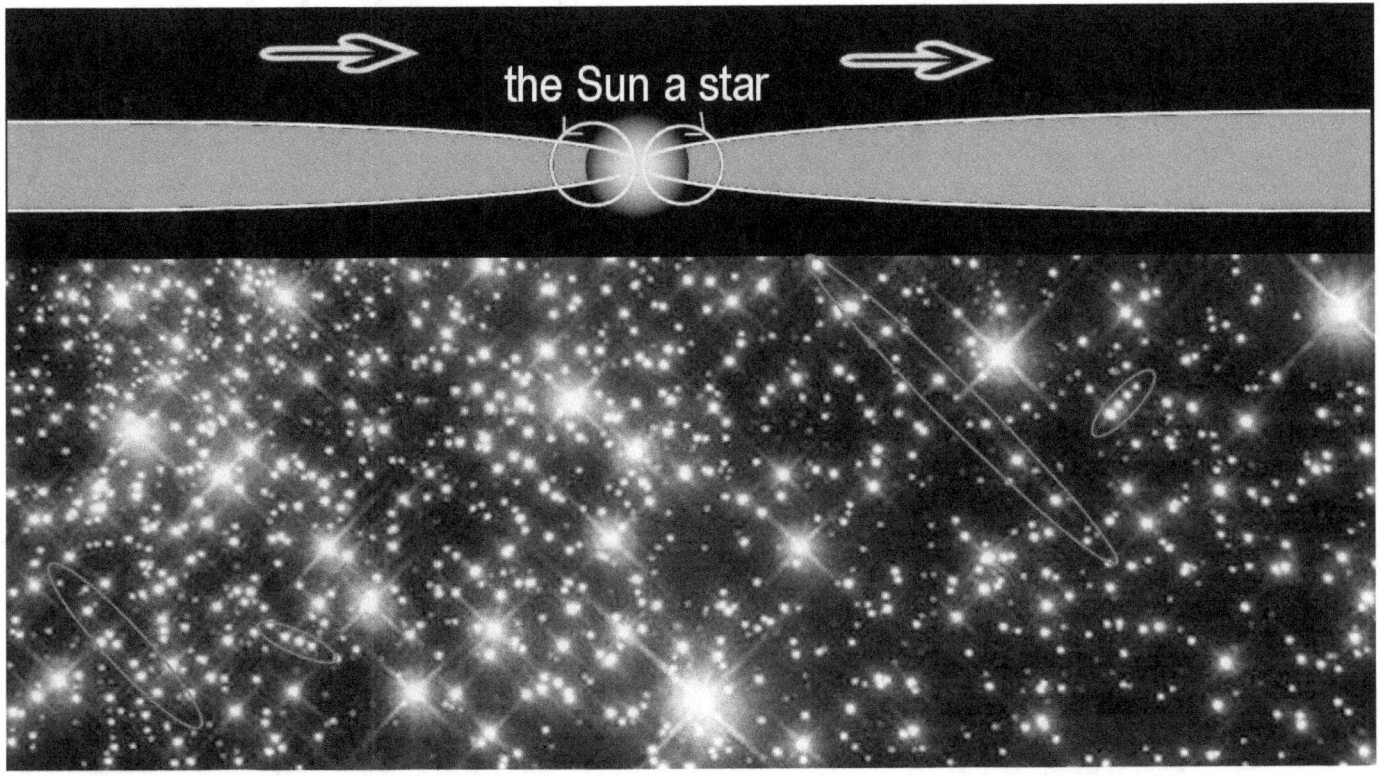

The universe is not consuming itself to produce energy. The universe is energy, and is producing everything with it.

## We fail aiming for fusion power

We fail, when we aim to operate outside the platform of the universe, aiming for fusion power for which no basis exists. Indeed, we do fail big time in every single one of our attempts to do this.

## Dead-end energy future

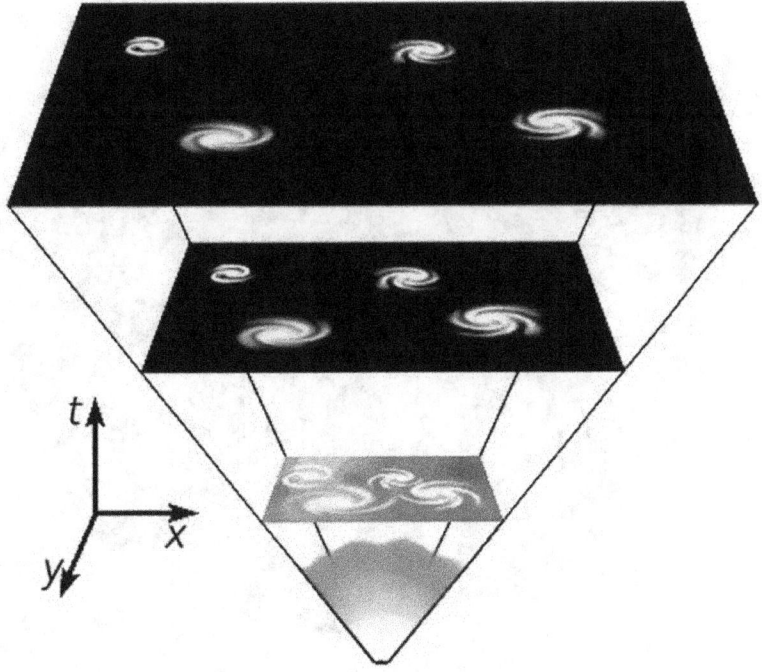

The point-source and dead-end energy future that the Big Bang theory parades as a concept, is false.

# Explosion is entropic and blows itself out

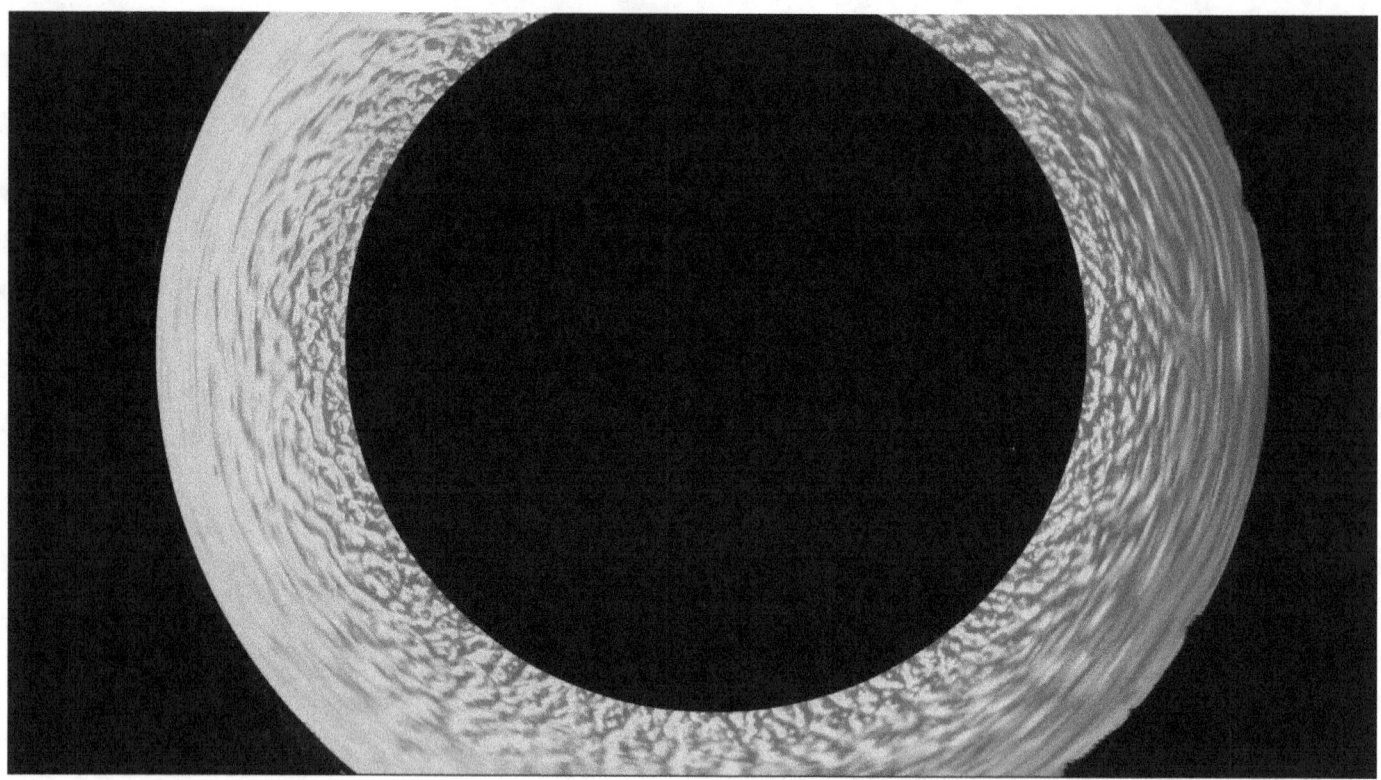

Every explosion is entropic and blows itself out. No evidence exists for it being the universal order. Every form of evidence reveals the theory to be a trap. We should celebrate that the theory is false, because if the Big Bang was real, humanity, would have no future.

## By clinging to the Big Bang trap

At the present time, by clinging to the Big Bang trap, the trap of universal entropy, humanity denies itself to have a future. It denies itself the universal energy resource, which is a resource that is not based on a consumable fuel, but is active everywhere.

# The anti-entropic energy resource

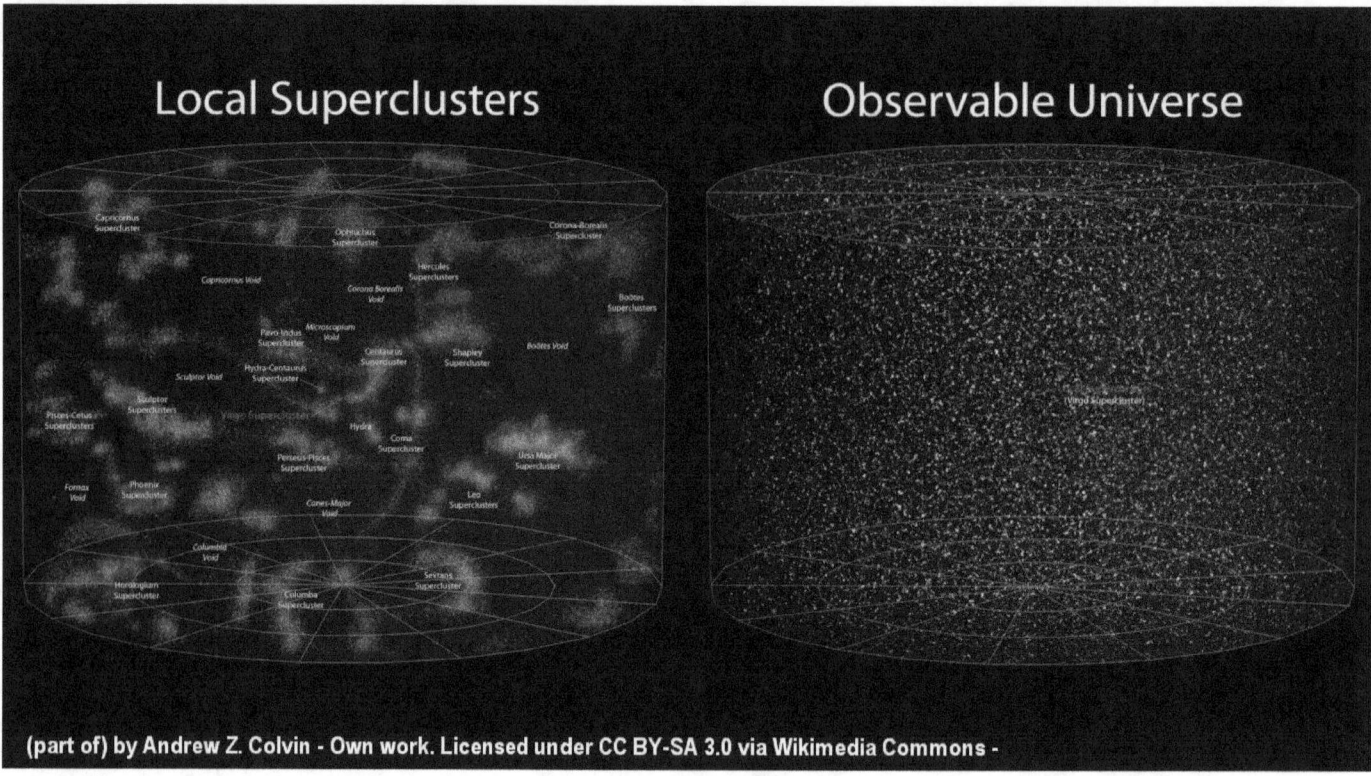

(part of) by Andrew Z. Colvin - Own work. Licensed under CC BY-SA 3.0 via Wikimedia Commons -

The anti-entropic energy resource that is rooted in the nature of the universe itself, is a resource that is constantly self-renewing. The cosmic energy resource is plasma - plasma in motion - it powers everything. Our future depends on tapping into this resource. Without it we have no future. As I said before, all of our gas, oil, coal, nuclear energy systems, are fuel-based systems that are inherently entropic systems. These entropic energy systems are depleted by energy use. They are dead-end systems. Only the cosmic-energy system is self-renewing and self-expanding.

## If the Big Bang was the truth

If the Big Bang was the truth, a nebula, such as the Crab nebula, would not exist. The crab nebula emits light energy at the equivalent of 75,000 Suns. Under the Big-Bang entropic theory, the light is the residual energy of a super-nova star explosion back in 1054 AD. However, no explosion, anywhere, has ever created a source of light that does not dim, but grows brighter. The light emitted by the Crab nebula, or any other nebula, is not residual energy from an explosion, but is light emitted by atoms activated with the movement of interstellar plasma that flows through the nebula, typically centered on a large sun.

## Plasma flowing through a nebula

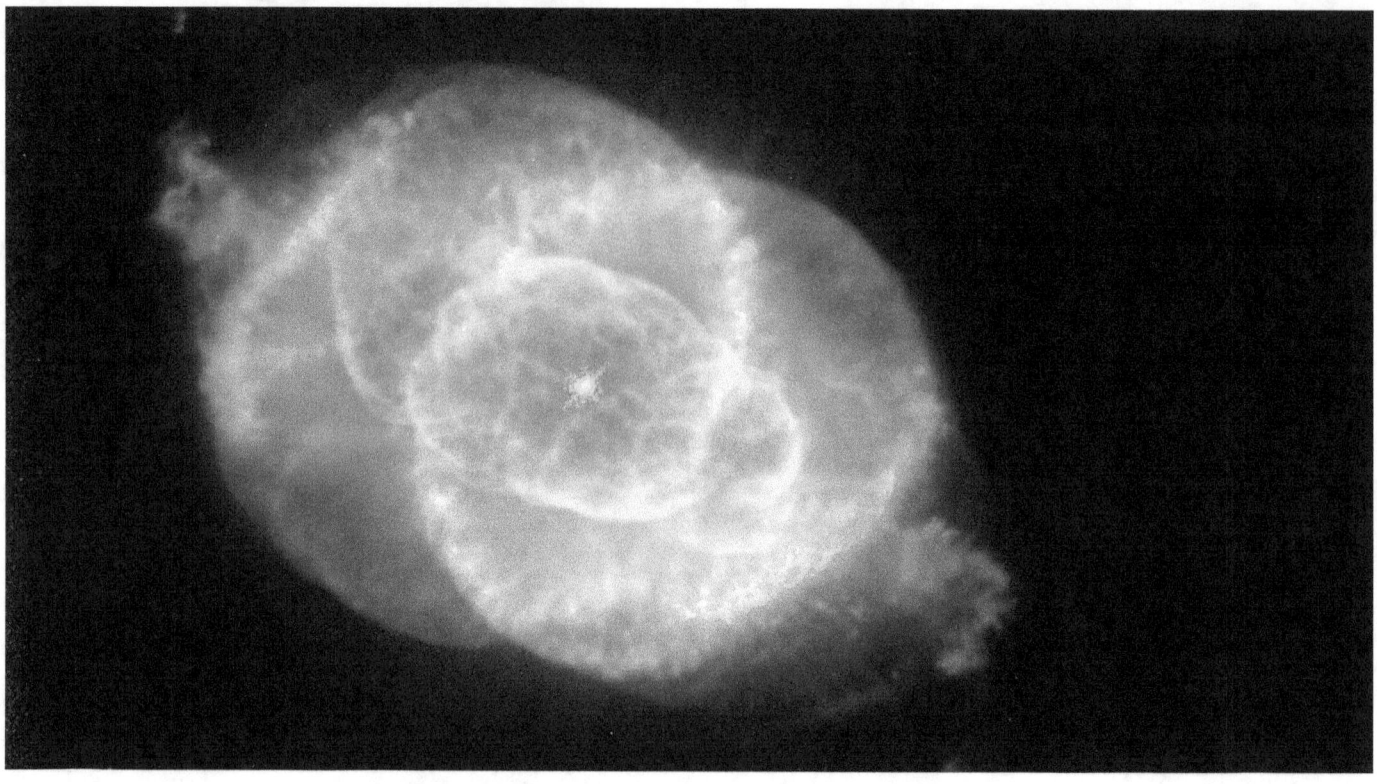

The flow-through process, of plasma flowing through a nebula, is indicated by the bipolar shape of nebulas that is often plainly visible.

## Our energy-future on Earth

In these types of focused plasma streams, our energy-future on Earth is located. The Earth orbits within a highly concentrated plasma flow that presently creates the solar plasma environment.

# The Red Square nebula

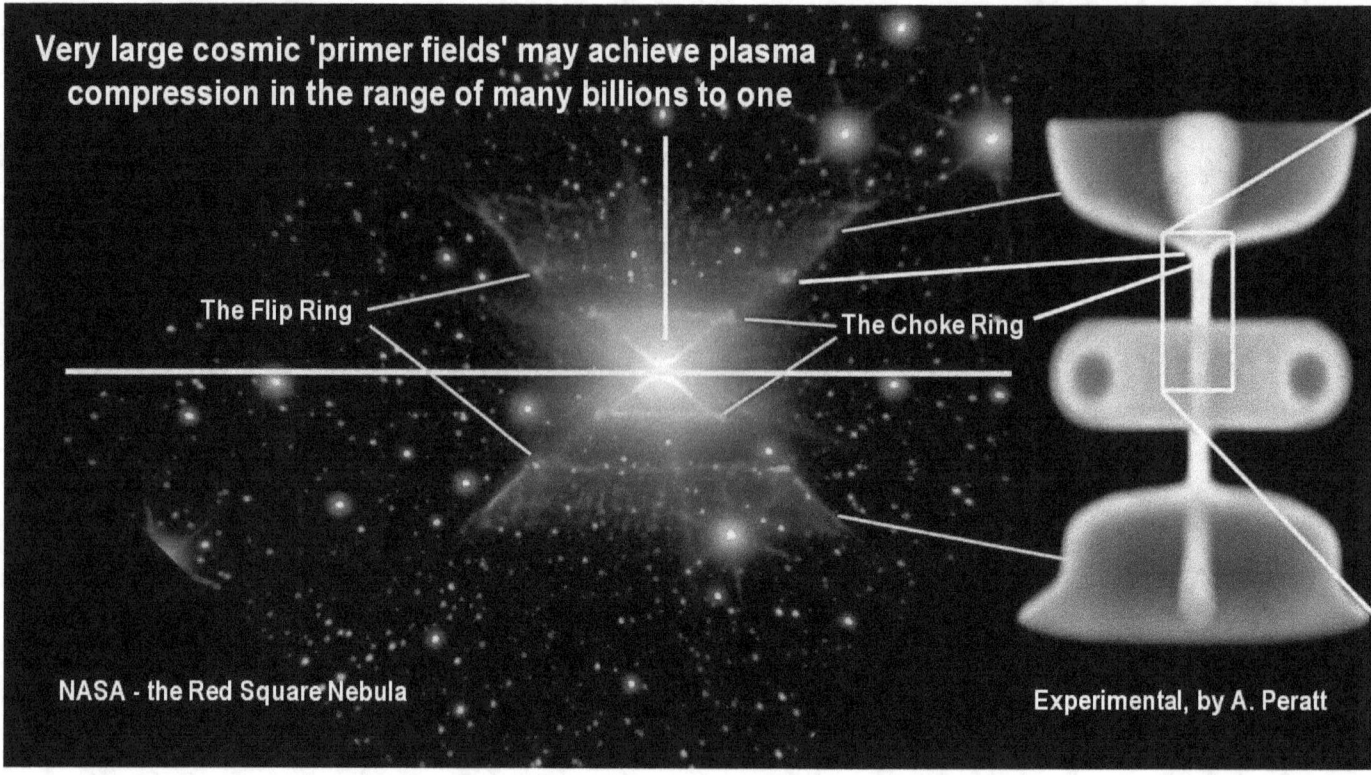

In some rare cases the shape of nebulas, such as in the case of the Red Square nebula, matches in principle the geometry observed in high-energy plasma experiments.

## Our interface with the stellar plasma streams

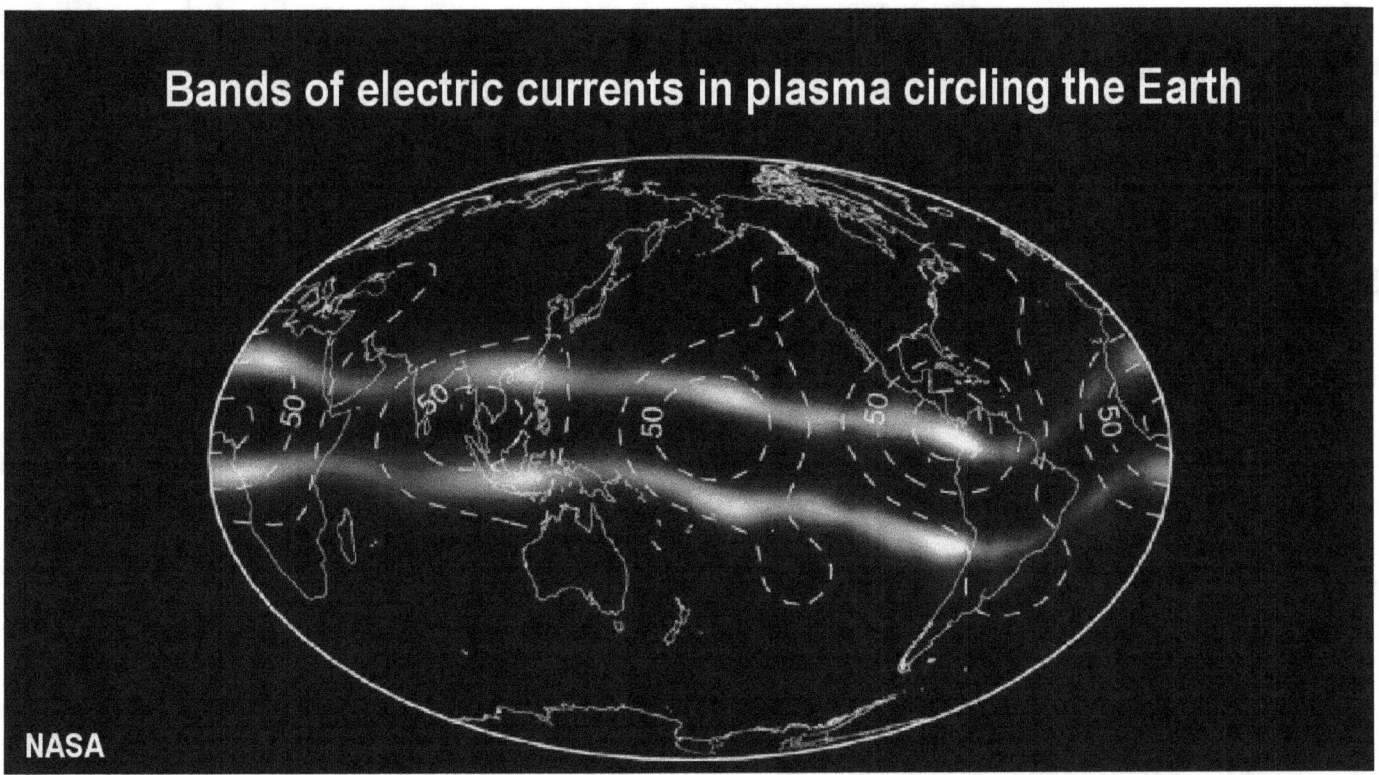

And as I said before, our interface with the stellar plasma streams, is visible in the form of equatorial plasma bands surrounding the Earth.

## Also visible on the Sun

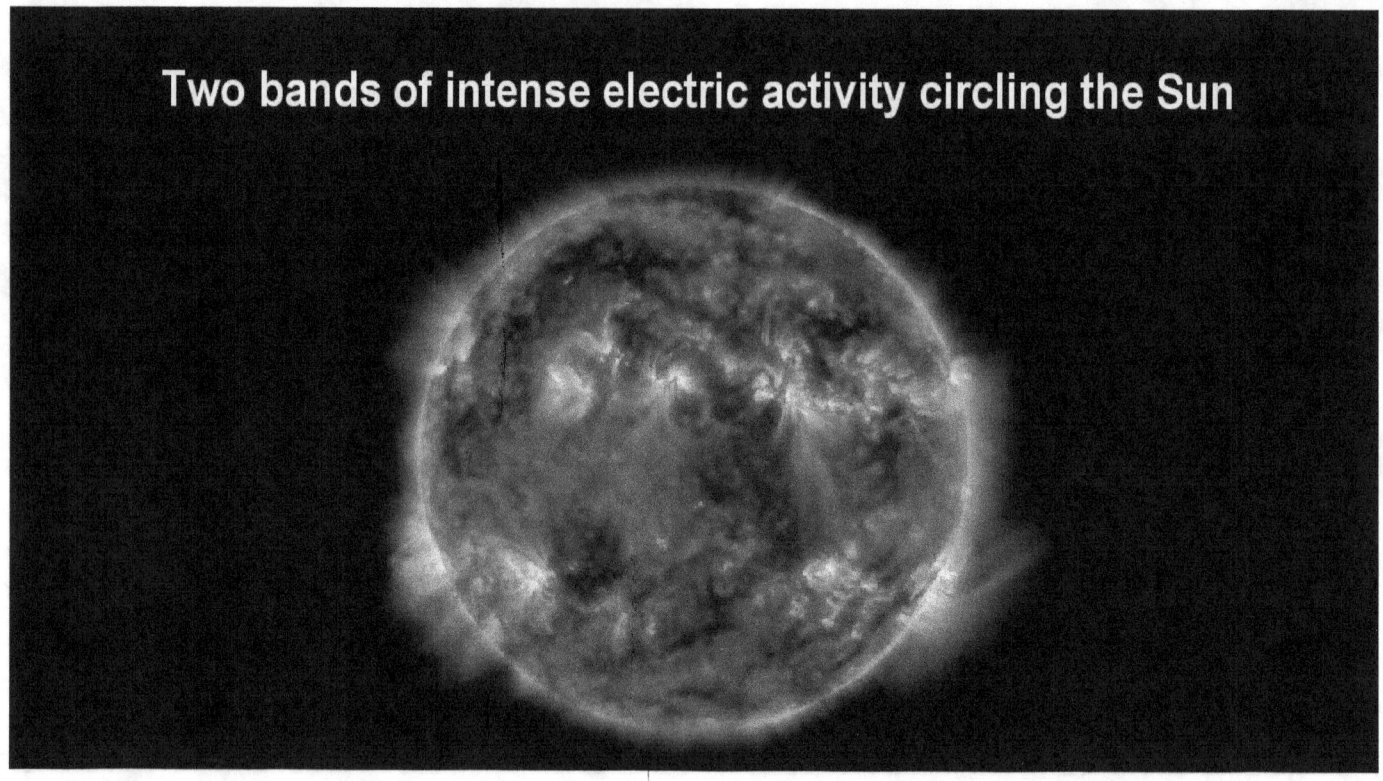

The same type of bands are also visible on the Sun that operates by the same electric principle, only much more powerfully so.

## The cosmic electric energy platform

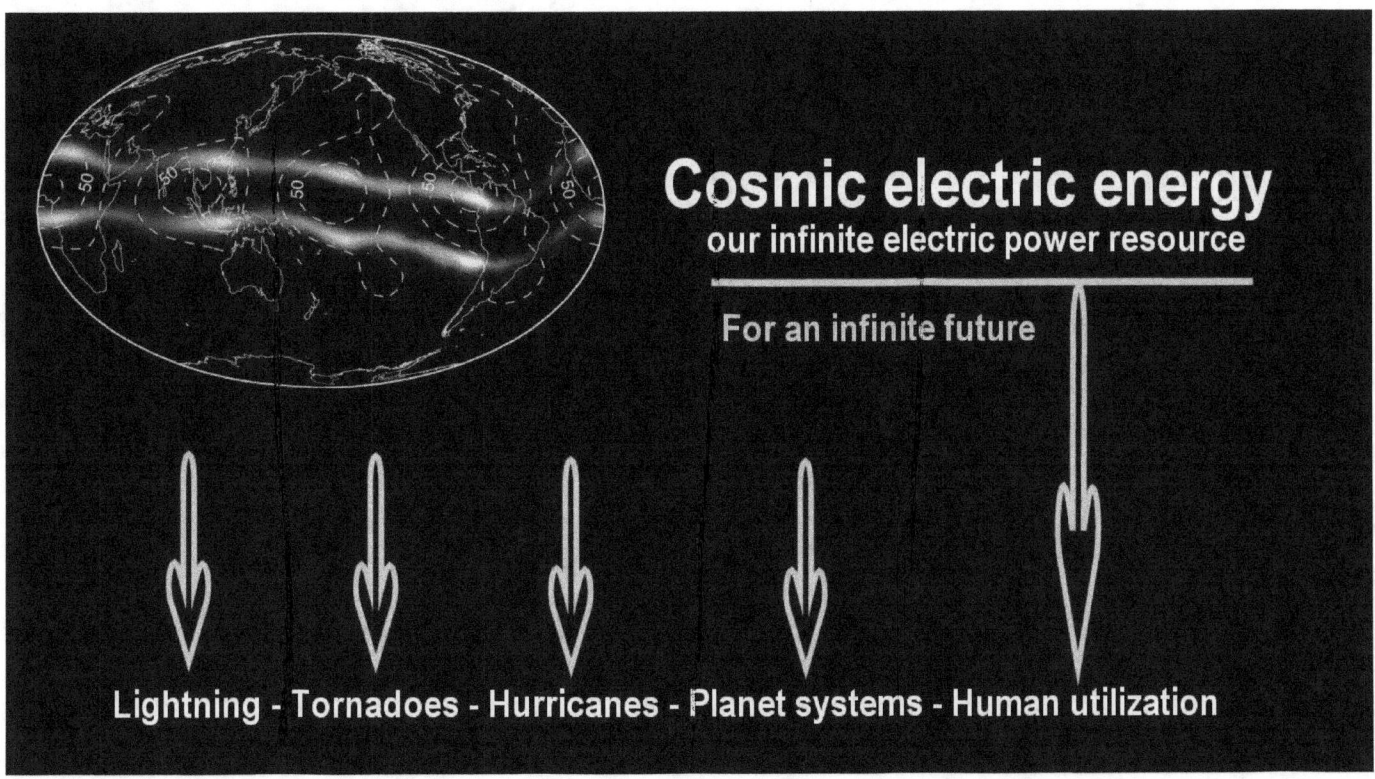

As I also mentioned before, the cosmic electric energy platform operates a number of large natural planetary systems, such as lightning, tornadoes, and the major global ocean and air currents. In comparison with the energy flux that powers these immensely powerful systems, our human energy needs are rather small.

It is this rather small additional cosmic energy utilization, that we will need to power our future with, which the Big Bang theory would deprive us of by its dream premise that plasma streams in cosmic space do not exist. By its premise, mass and gravity are the only forces that are recognized to exist as a basis for all effects.

# The Big Bang Suicide Pact

"The Big Bang Suicide Pact"

# The Big Bang theory is a trap

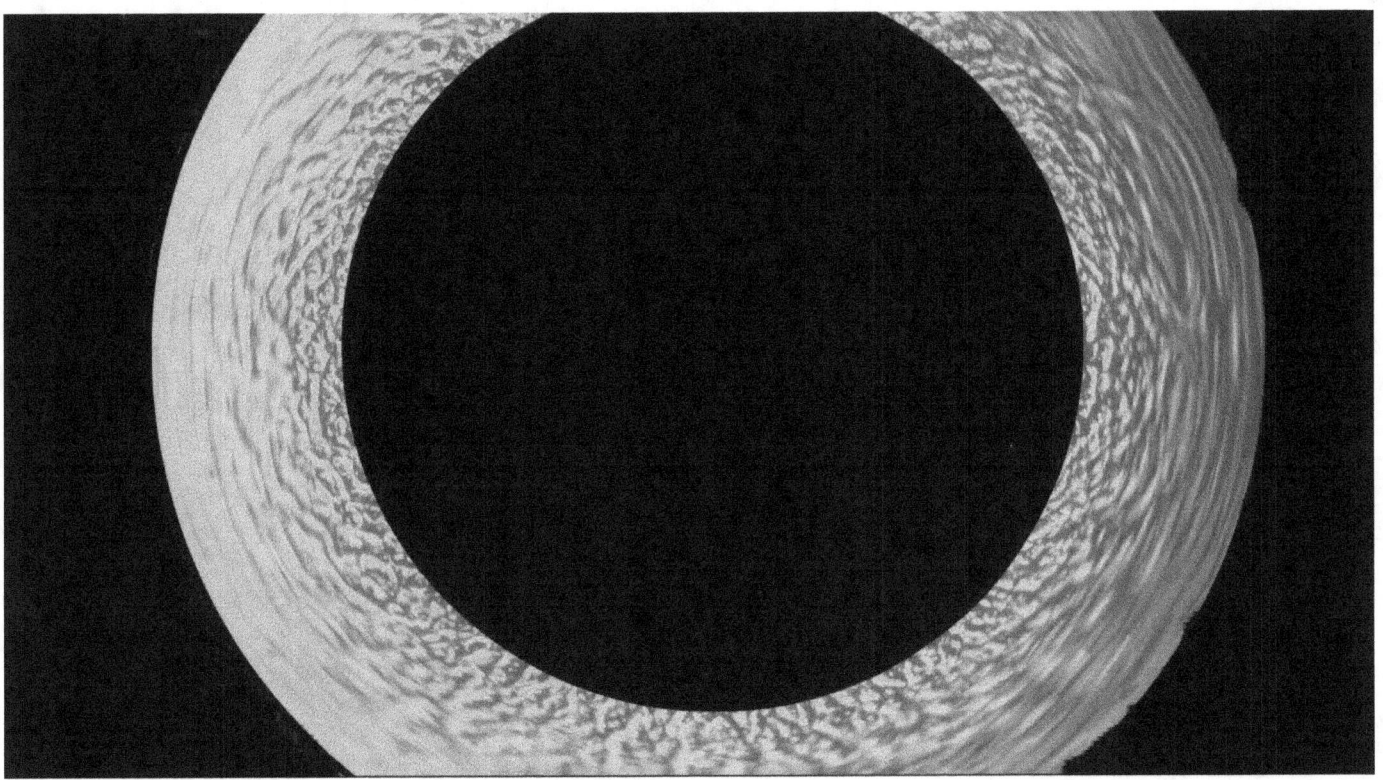

The Big Bang theory is a trap that is empty, a ring of smoke without substance. We always come back to this form as a model for entropy.

In the small-minded prison of universal entropy, the Big Bang theory acts as a global suicide contract that enforces energy starvation, and all kinds of related forms of starvation.

# The biofuels genocide contract

The western imperial system is presently murdering 100 million people a year with the biofuels genocide contract that demands that vast quantities of high-value food are being burned as fuel in automobiles in a world that has a billion people living in chronic starvation.

# The consequences of the difference

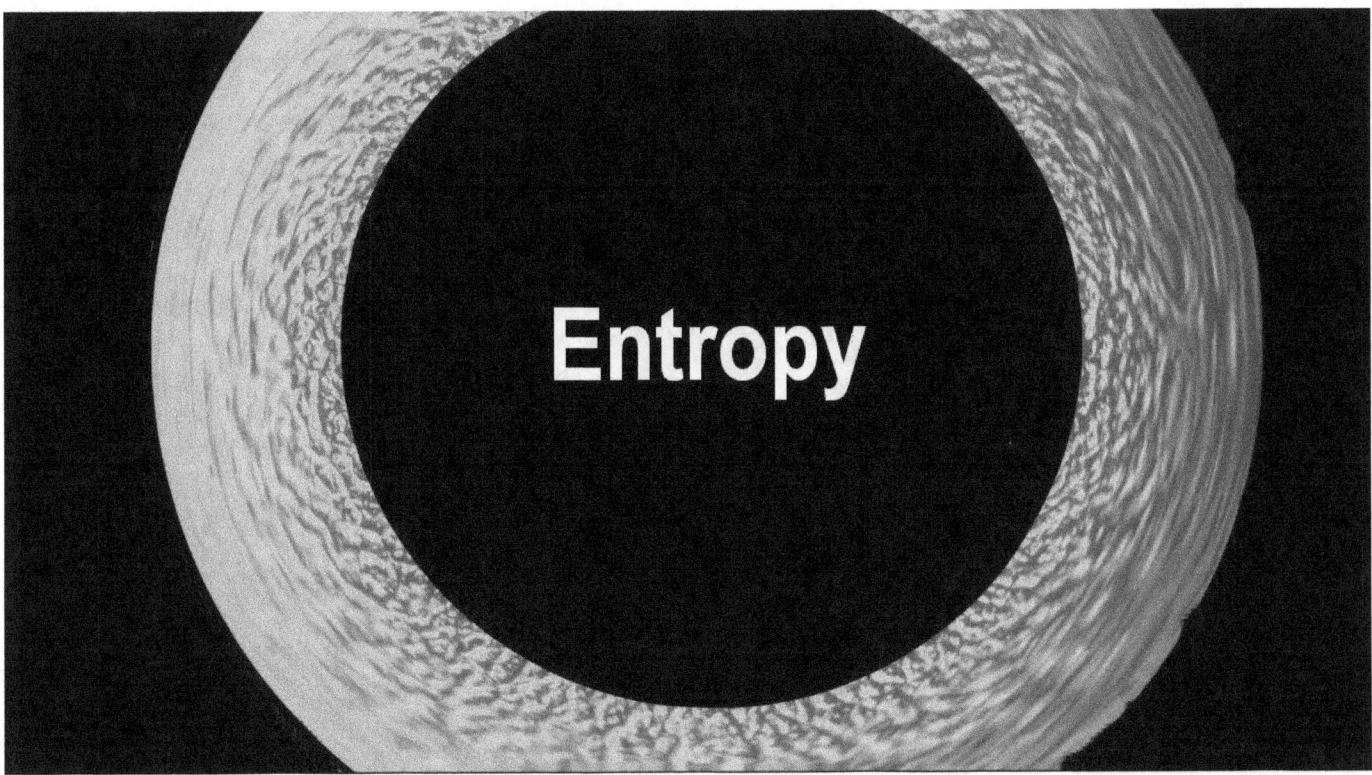

The consequences of the difference between the Big Bang cosmology, and the plasma cosmology that the Big Bang is designed to obscure, renders the controversy a deadly affair, which goes far beyond being merely an academic issue. The difference between the Big Bang system of entropy, and universal anti-entropy, is so vast that it has become an existential issue on the global scale. If the Big Bang Cosmology is not aborted, much of humanity will die in the near future from the imposed energy starvation and related insanities, all as the result of the assumed entropy of the system that society has been taught that it is bound to.

## In this trap, the Sun is deemed entropic

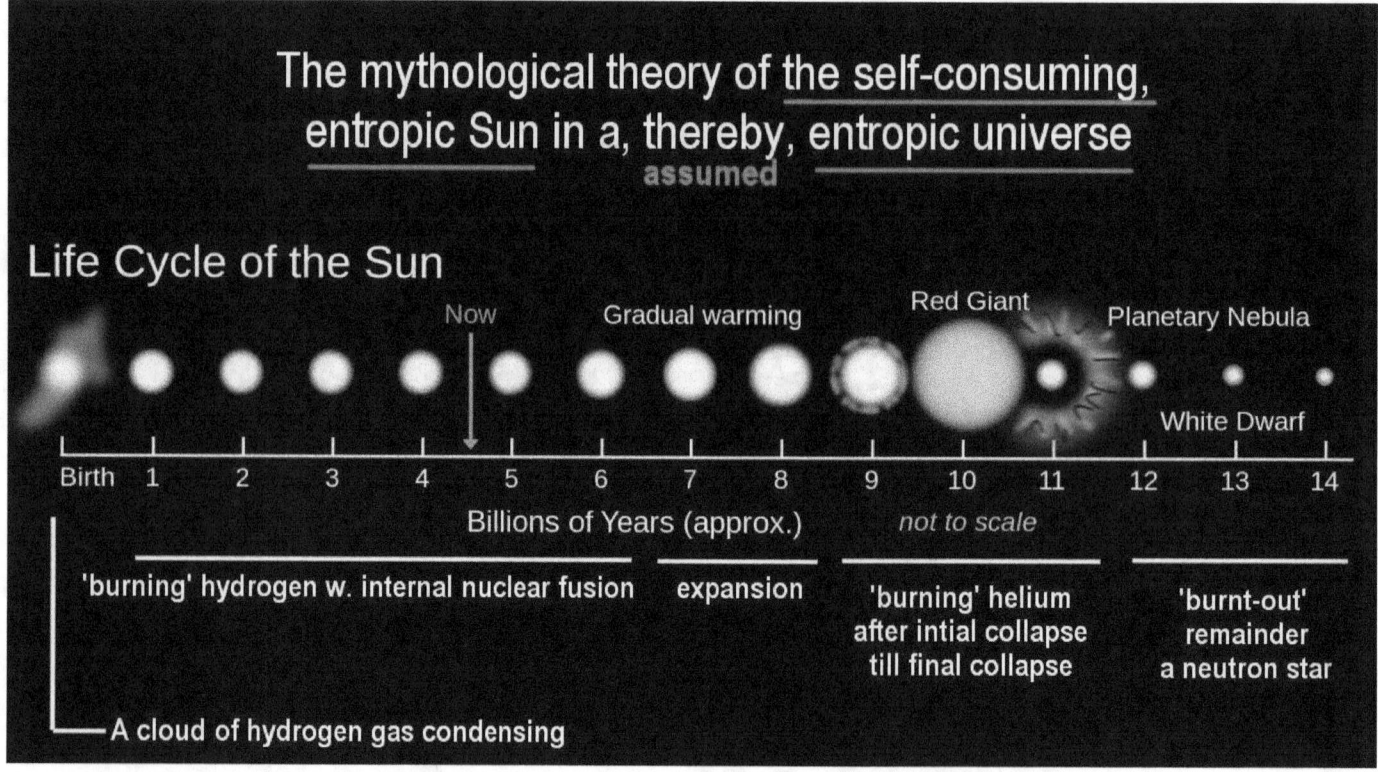

Because the Big Bang universe is deemed to have evolved from gravitational condensation of primordial dust and gases into stars, planets, and galaxies, no energy resource is deemed to exist that maintains anything. In this trap, the Sun is seen standing by itself, deemed entropic and burning itself up, for the lack of recognition of the cosmic energy resource.

# This dying-star solar model

> ## The most important issue of our time
> ## Big Bang versus Plasma cosmology
>
> | | |
> |---|---|
> | The universe evolved from the gravitational condensation of primordial dust and gases into stars, planets, and galaxies. | The universe is 99.999% plasma, Every sun is made of plasma, and is powered by plasma fusion with the electromagnetic force. |
> | The Sun is entropic, consuming hydrogen as a fuel in internal fusion, until the fuel is depleted, whereby the Sun dies. | Planets are formed from atoms synthesized by the Sun with the fusion of plasma that is dynamically drawn from cosmic space that cannot be depleted. |
> | Every energy resource on Earth is entropic. When the resource is depleted, humanity dies. | Cosmic plasma, as our energy resource on Earth, is self-renewing. With it, humanity has an infinite, energy-rich, future. |

In this trap, the Sun is assumed to be an isolated entity that is burning the hydrogen atoms that it is deemed to be made of, burning them as a fuel in internal nuclear fusion processes, until the fuel is depleted, whereby it dies in numerous ways. This dying-star solar model is then applied to the terrestrial energy resource model, which renders our future subject to depletion. Little hope remains after this. When the Earth's energy resources are depleted, and we are getting close to that, humanity has no option under this model but to die as a consequence, because no external energy resource is deemed to exist that would keep humanity going forever.

Now compare this inherent death spiral, with its opposite, the plasma cosmology. Here the universe is made up of plasma. This means that 99.999% of the universe, as plasma, carries an electric charge, and thereby electric energy. This means that a sun is not made up of cosmic dust and gases, but is made up of plasma with an electric quality, whereby it becomes electrically powered with a process of interstellar plasma becoming drawn to it that becomes fused into atoms on its surface with the electromagnetic force generated by the movement of plasma itself. On this platform all planets are formed from atoms synthesized by the Sun. Since plasma in cosmic space is self-renewing, it cannot be depleted. Consequently a sun cannot die. Neither can the Earth ever be deprived of an energy resource, because the cosmic plasma that we can tap into, is not a fuel that can be drawn to depletion, but is an enduring quality of the universe. With it, humanity has an infinite, energy-rich, future.

# For as long as the Big Bang cosmology rules

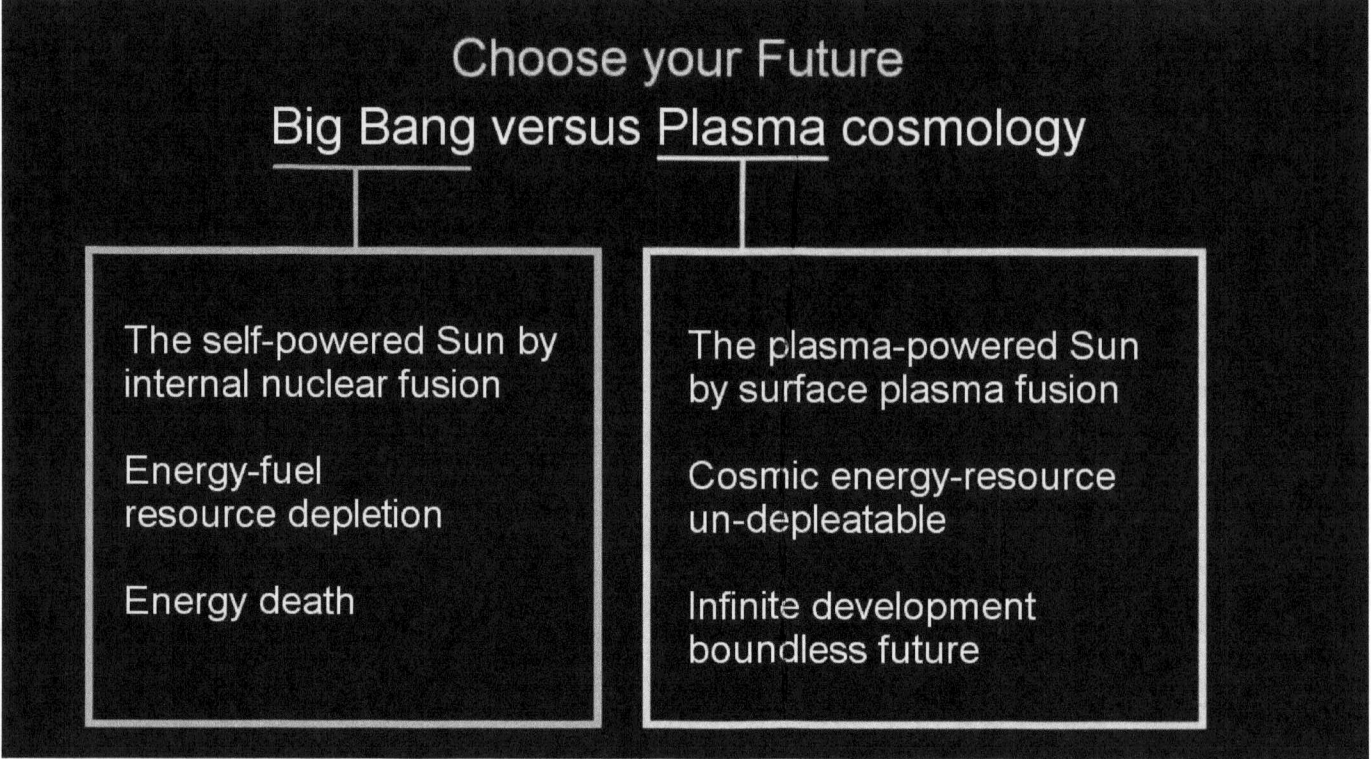

This means we have a choice before us. For as long as the Big Bang cosmology rules in the mind, humanity is effectively trapped by it, to die the death of the assumed energy resource depletion. This belief is apparently so strong that plans are promoted to harvest heliom-3 from the moon as a fusion-energy fuels, in spite of the fact that the last experiment required a million times greater energy input to cause the helion-3 fusion to happen, than it gave back as released energy. That's how heavily the assumption of energy entropy weighs on humanity, that it looks for miracles to circumvent it, while remaining unaware of the anti-entropic energy option that the Big Bang theory obscures.

The Big Bang entropic cosmology operates as this type of deadly package that includes numerous similar aspects.

Inversely, in the Plasma Universe, that stands as the total opposite of the entropic theory, everything is recognized to be actively powered by the forever-flowing cosmic, plasma energy streams. The Plasma Universe, too, operates as a package. This package includes the plasma Sun, with cosmic plasma fusion occurring on its surface. It promises humanity an energy resource to tap into, that cannot be depleted, whereby humanity has an endless energy-rich future.

This leaves us with the question: which package will determine our future? The Big Bang package offers death by depletion of everything. The Plasma Universe package offers unlimited resources for development, life and abundance. Which of the two option would you choose as a basis for building a civilization on?

# The Plasma Universe offers itself as an open door

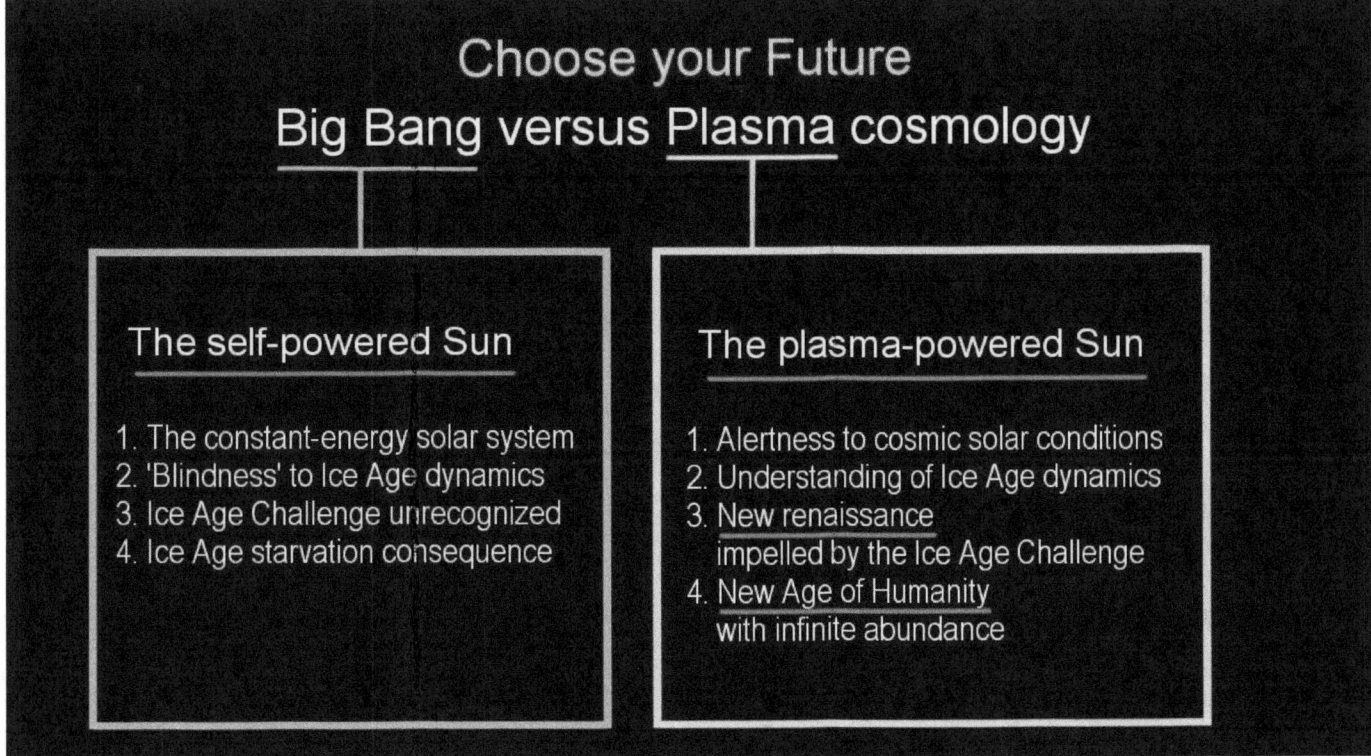

The Big Bang offers a trap. The trap has numerous faces. Each face reflects the notion of the self-powered Sun, whereby it hides the coming Ice Age and its consequences.

In contrast, the Plasma Universe offers itself as an open door. The open door has likewise many faces, with each reflecting the anti-entropic plasma-powered Sun and its electric principles that determine the Ice Age dynamics, which are knowable and enable us to respond to the dynamics to protect our civilization.

# The Plasma Universe

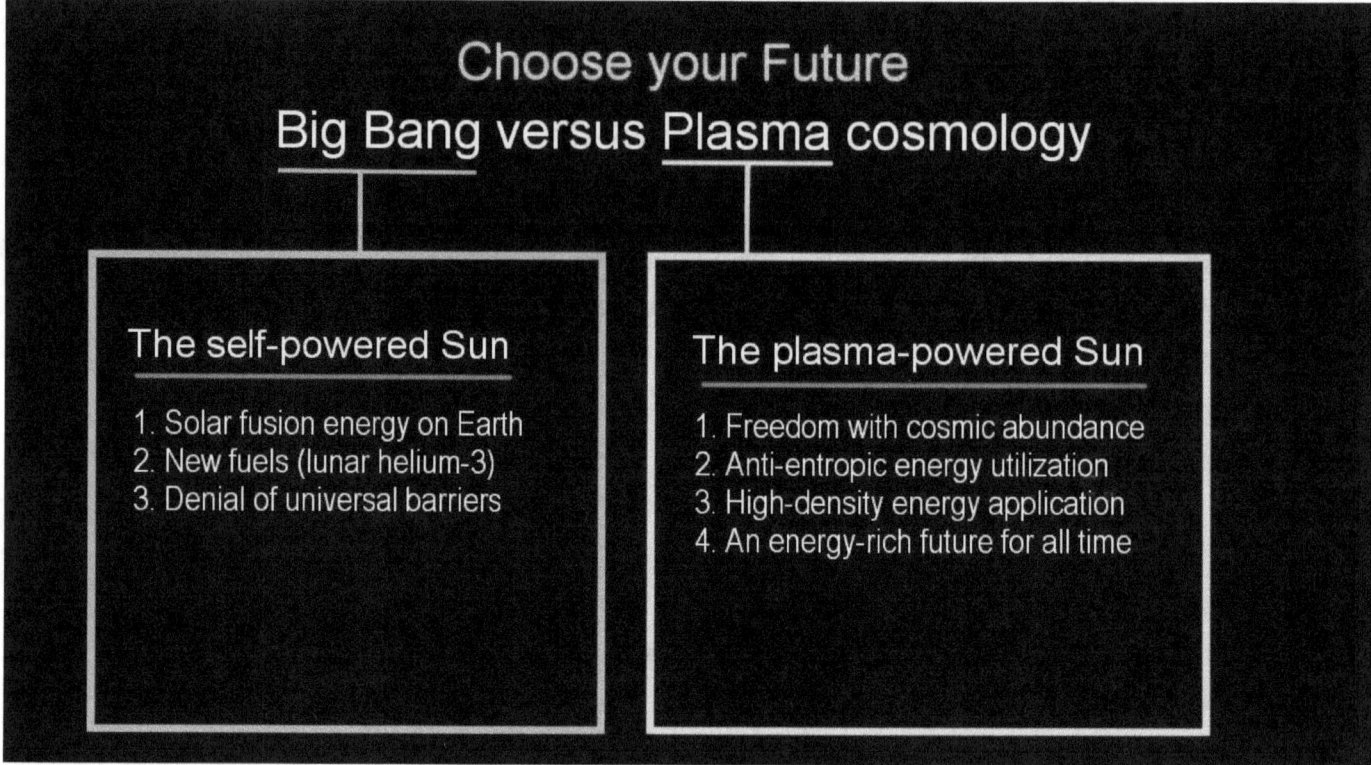

The Big Bang theory includes a number of entropic elements that set up traps in the mind that inspires desperate measures:

1. The solar fusion energy dream on Earth is a dead-end trap, because the Sun is not powered by internal atomic fusion that society attempts to replicate.

2. Since fusion power doesn't work, out of desperation, society now looks to the moon to mine helium-3 from its dust, in the desperate hope that helium-3 will be a useful fuel, though it. is the hardest element to fuse.

3. The desperation causes a denial in science, of the universal barriers that the universe has erected against atomic fusion to protect its integrity.

The resulting entropy-inspired atomic-fusion energy hope stands nevertheless as but a brilliant dream with an empty center.

The Plasma Universe, in comparison, offers humanity real ready-made energy without the need for burning any type of fuel. It offers humanity

1. A great energy-freedom with cosmic abundance

2. It enables anti-entropic energy utilization

3. It enables high-density energy applications

4. It opens the door to an energy-rich future for all time to come

In the real universe, the plasma universe, the Earth is afloat in a sea that is energy, which powers the Sun. Why then would we need to 'produce' energy?

# Segment 2 - Entropy - Empire – Economic

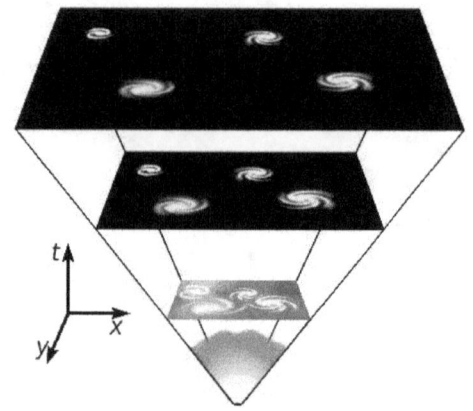

# Big Bang Blow-Out

## Segment 2    Entropy - Empire - Economics

The Big Bang Cosmology would have the universe hyper-explosively expanding, even while it is winding down towards its inherent energy depletion death. It has become the model for an equivalent economic system, if one can call it that. The result is the system of empire that spreads itself across the world, explosively, as an engine to acquire wealth by looting instead of creating, which promptly collapses economies and thereby the value of the stolen loot that stands as a claim against a dying economy.

Since empires exist by stealing, empire drives for war. All wars in history have been instigated by the forces of empire for the purpose of stealing. This is the only platform that empires that empire can exist on. The platform is destructive, as stealing and wars inherently are. The platform is thereby entropic, and as a consequence self-collapsing. All empires have collapsed on this platform, and with them the societies that have allowed themselves to be dominated by them.

Empire's cycle of entropy has been subsequently projected onto the universe, with the Big Bang theory to give empire a legitimate face.

# The Big Bang trap is paraded as brilliant

### Entropy / Empire / Doom

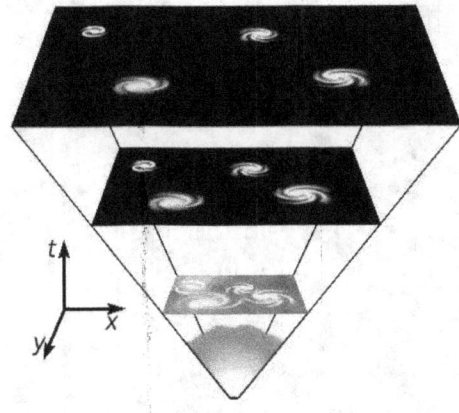

After the Big Bang start-up
the universe is winding down

### Anti-entropy / Principle / Development

A universe that is forever 'beginning'
self-powering, self-expanding without end

The Big Bang trap is paraded as brilliant, even while it is an illusion that has no substance at its center. However, the Big Bang theory accurately defines the nature of empire. With this theory, the system of empire projects its own platform onto the universe, as if to give its empty platform the appearance of legitimacy, and its inevitable collapse the face of a natural process, which it says is fundamental to the universe itself.

## With the Big Bang theory

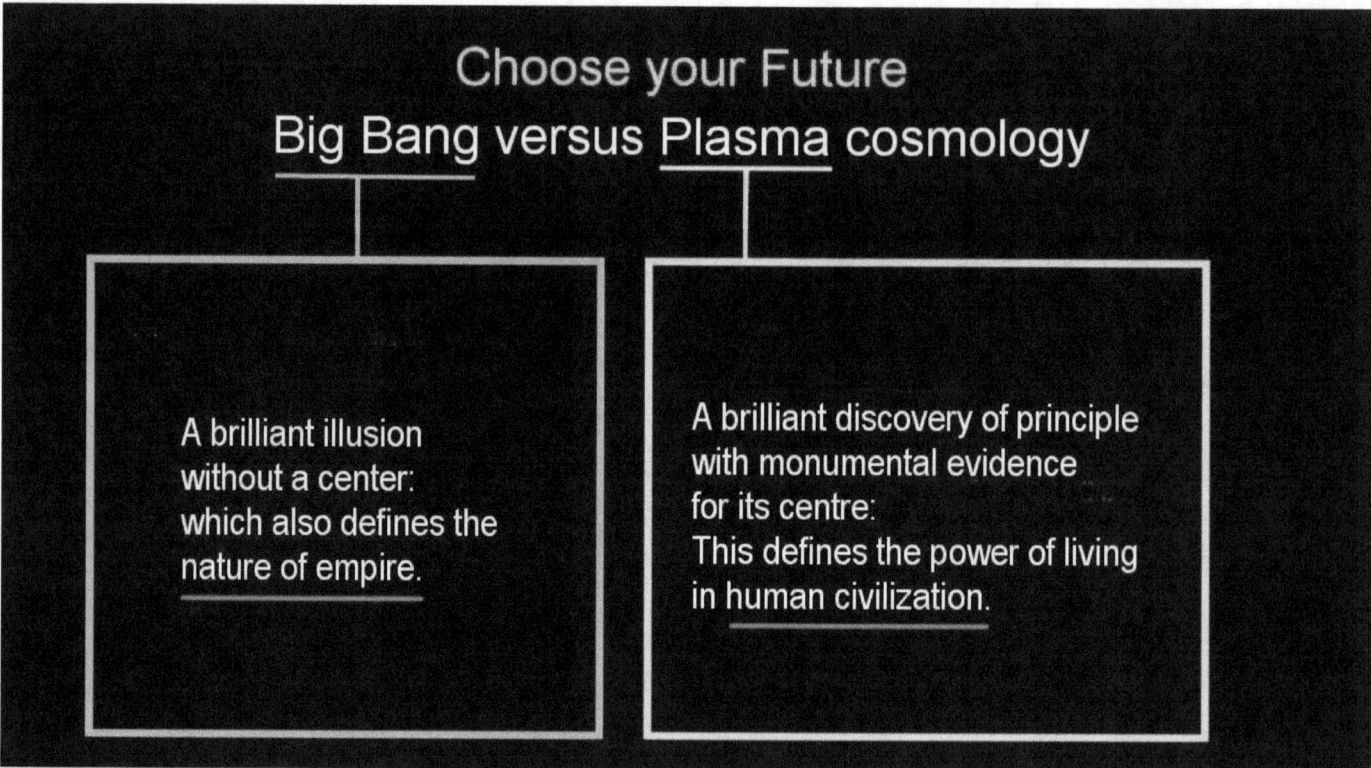

With the Big Bang theory empire says in essence to a struggling humanity, 'the collapse of your civilization is natural. What winds up, winds down again. There is nothing you can do about it. You are impotent, so don't even bother to try. Just look at us: Every empire that ever was, has collapsed. Nobody has yet succeeded in preventing the collapse of empire at any point in history, and of course the collapse of civilization with it. That's built into the nature of the universe. Nobody can alter that. So, just accept the inevitable and lay yourself down to die." This appears to be the modern song.

## The lyrics for the empire-song are a lie

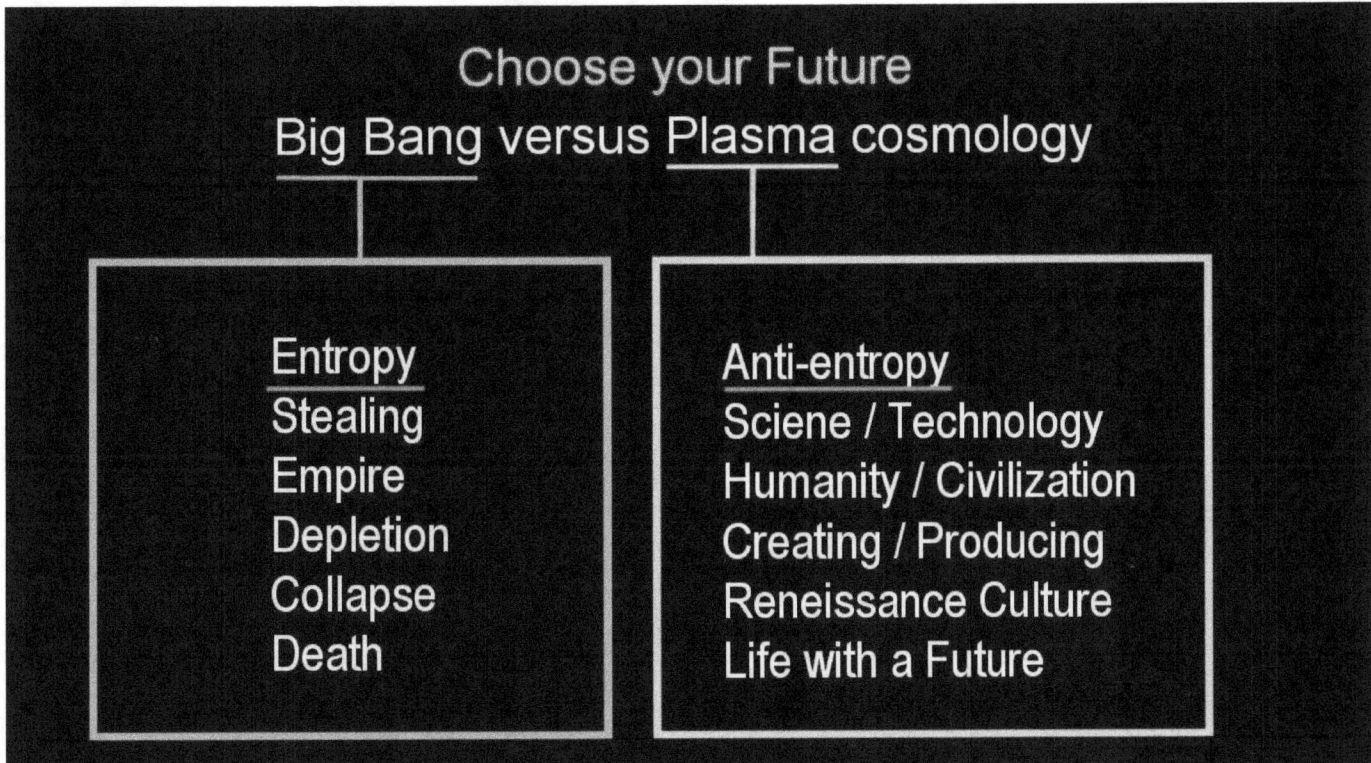

Fortunately, the lyrics for the empire-song are a lie. While it is true that every empire in the entire history of humanity, has collapsed itself without fail, it is also true that every collapse of empire resulted from its inherent emptiness, within, from its devotion to the entropic dream of gaining riches by stealing. In order to prevent its doom, empire seeks to control the world, to keep humanity small so that it will not rebel, but tolerate the pains of entropy, such as providing evermore bailouts to shore-up the thievery.

With these sayings, empire lies to society, big time.

# While no empire has ever survived

While no empire has ever survived its defective entropic dream of gaining riches by stealing, whether by trickery, war, or slavery, humanity has been successful a number of times in different places, at evading the resulting doom for its civilization, by human destruction under the system of empire.

# Society has freed itself and its future

Society has freed itself and its future, by unlatching itself from empire and its 'corrosive' platform of universal entropy, with a commitment to the general welfare of the nation and nations.

# By placing itself onto the platform of anti-entropy

Society has freed its future by placing itself onto the platform of anti-entropy, the platform of the creative renaissance of human development, the platform of cultural, scientific, and technological progress.

# On this platform nothing is winding down

The National Centre for the Performing Arts - Beijing, China (2007) seats 5,452 people in three halls by French architect Paul Andreu
images Wikimedia Commons, CC BY-SA 3.0 and GFDL

by Fanghong
by Evilbish

On this platform nothing is winding down, except the existence of empire that by its very nature has no place to exist on an anti-entropic platform.

# Empire lacks the intention to be creative

Empire lacks the intention to be creative, with its focus on stealing. Empire has never produced an iota of good for the general welfare of humanity. It enforces thereby its own end.

# The Big Bang theory needs to be dealt with

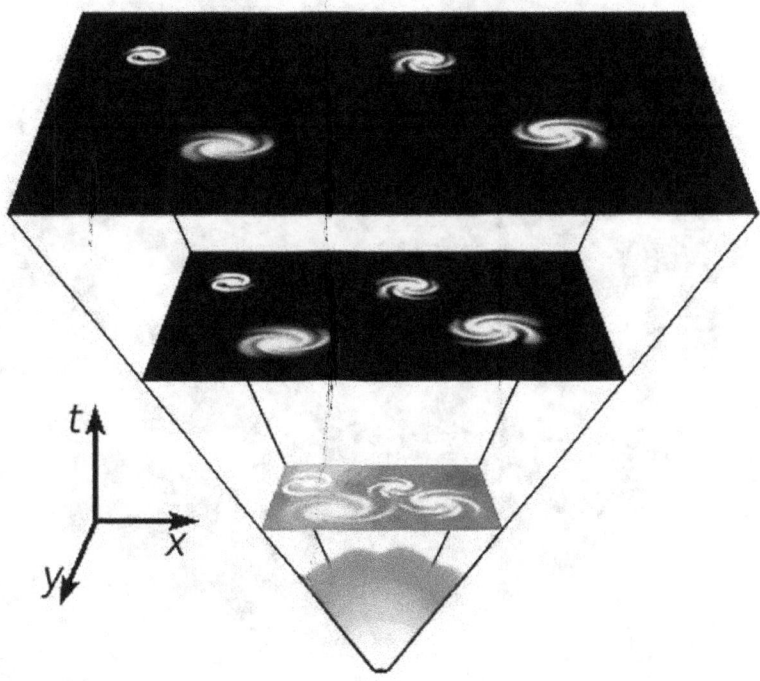

The Big Bang theory needs to be dealt with for humanity to unlatch itself from the entropy of empire, because the theory projects the self-collapsing nature of the system of empire unto the universe. The Big Bang theory is a false theory. It is a fairytale that has no foundation. It invites society to dream the dream of entropy, the dream of universal depletion.

# Theory of the inner emptiness of the universe

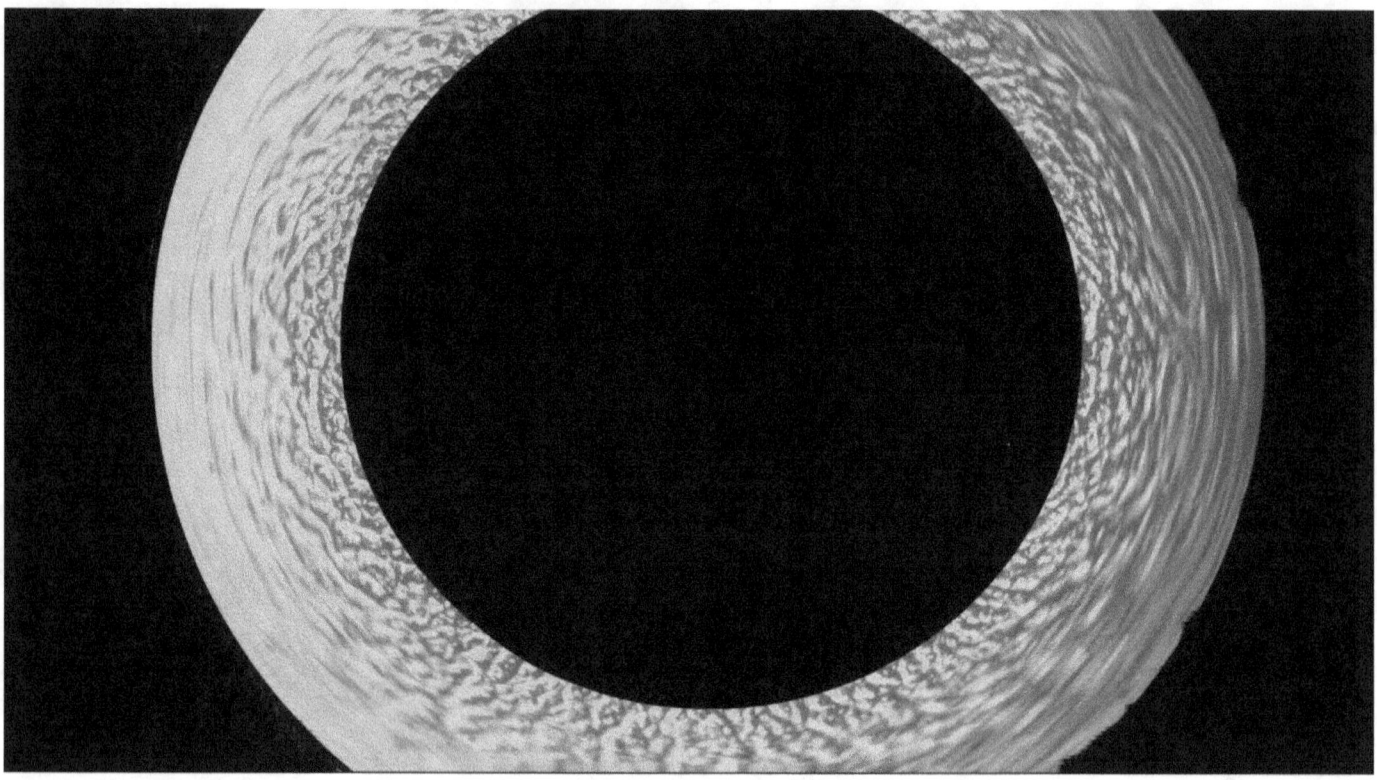

The Big Bang theory, the theory of the inner emptiness of the universe, which it projects unto the universe as if it was real, has the effect that it blocks its opposite, the Plasma Universe, which is real, from becoming recognized. That's where its danger lies. It seeds emptiness, where there is in reality boundless energy in every respect.

# The Big Bang philosophy

Entropy / Empire / Doom

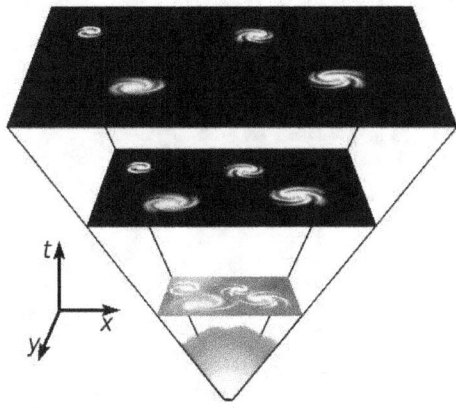

After the Big Bang start-up
the universe is winding down

Anti-entropy / Principle / Development

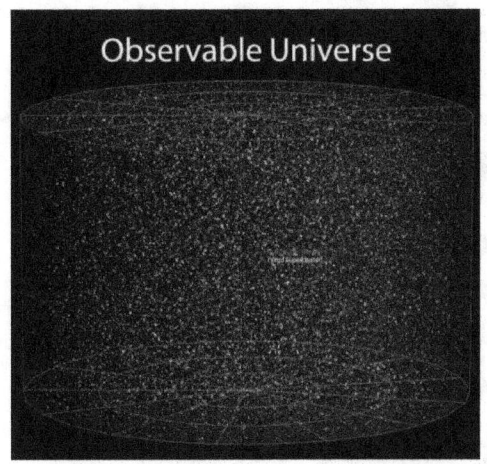

A universe that is forever 'beginning'
self-powering, self-expanding without end

The Big Bang philosophy blocks the brilliant discovery of universal principle for which monumental evidence does exist. By society's understanding that the Big Bang theory is fundamentally false, and that its philosophy of entropy is but a tragically empty, false dream, society unlatches itself from the deadening consequences of the dreaming.

## Humanity becomes free

By unlatching itself from false dreaming, humanity becomes free to accept the brilliant discovery of the principle of anti-entropy.

# The principle of cosmic anti-entropy

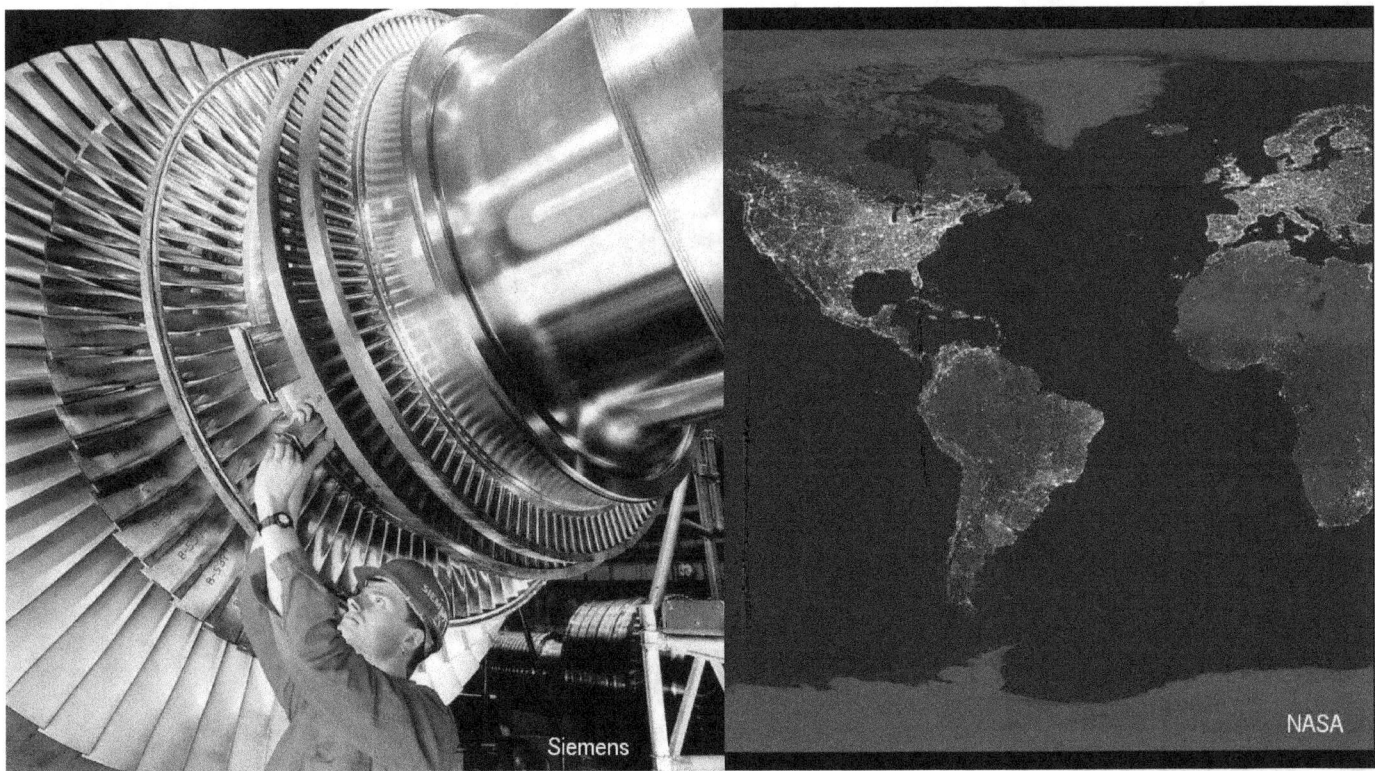

The principle of cosmic anti-entropy is also humanity's own inherent anti-entropy, for which monumental evidence exists that defines the power in intelligent living in a human civilization, which becomes applied inherently scientifically, and progressively with technology, and of course expansively in anti-entropic economics, such as with high-temperature, automated, industrial processes utilizing advanced materials and processes.

# The Big Bang theory blocks humanity

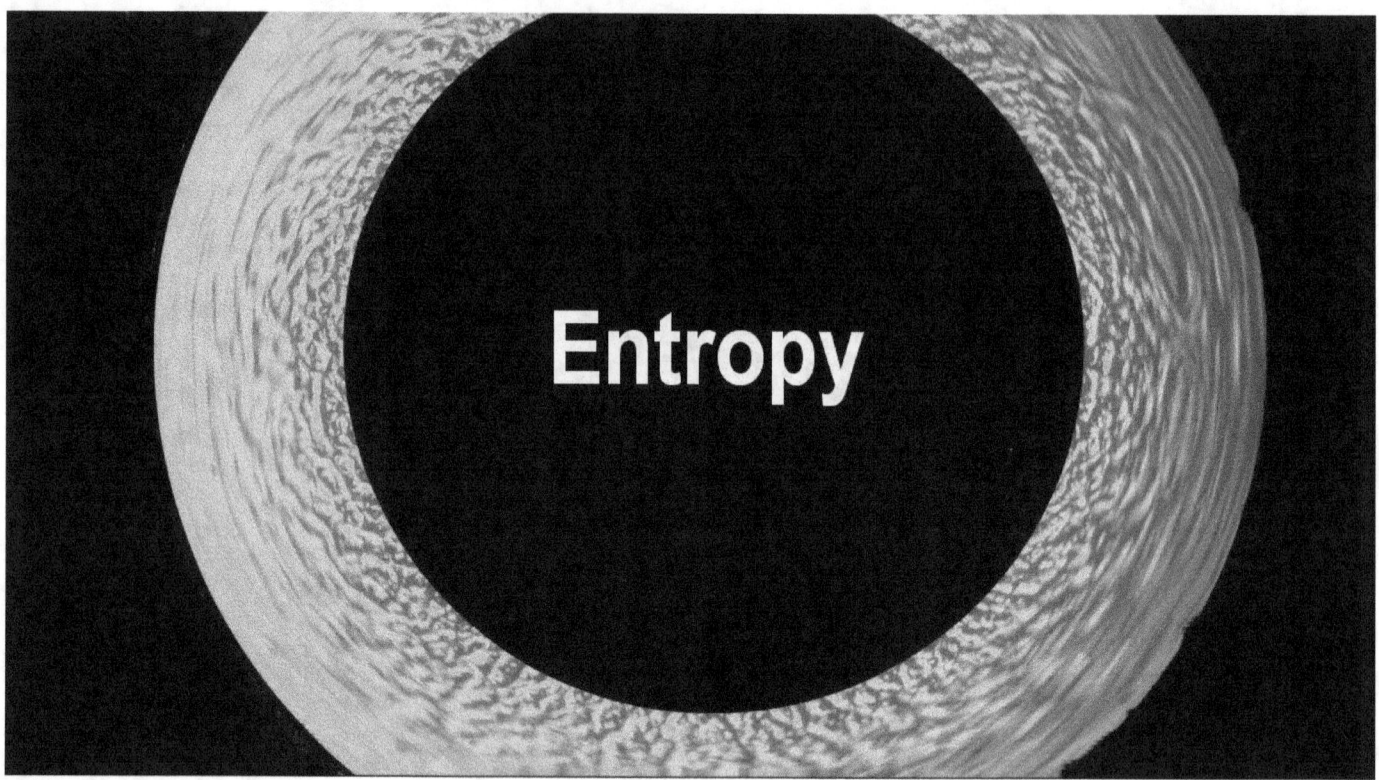

The Big Bang theory blocks humanity from being human. It is a global suicide contract that has been cleverly dreamed up for the masters of empire, as a tool that closes the door on humanity's progressing beyond the empty entropic base that empire dwells on, the entropic base of fuel-based energy production, privately owned, and privately exploited. The Big Bang theory may have been developed in part for the purpose of keeping humanity empty inside and without hope.

## The plasma universe concept

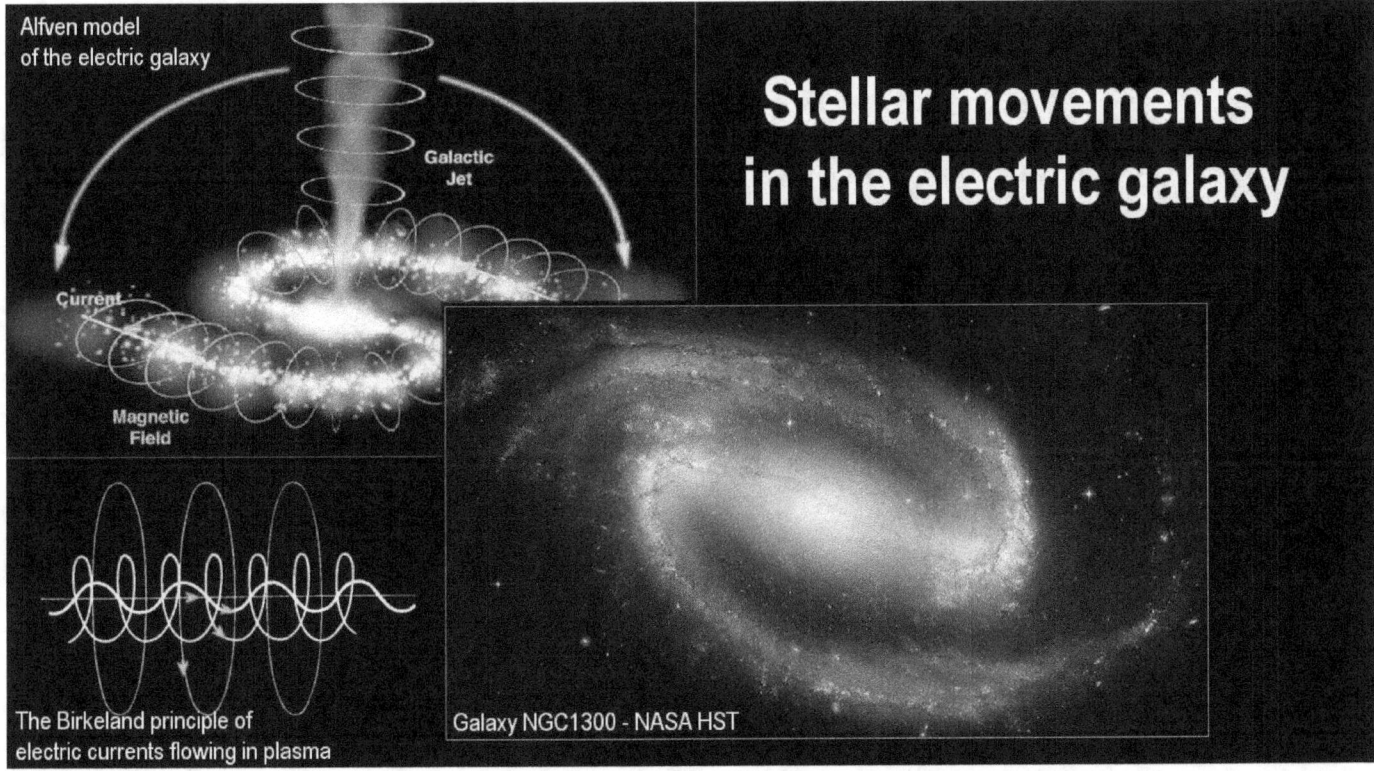

The opposite concept, the universally powered concept, the plasma universe concept, was pioneered in the late 1800s by the Swedish electrical engineer Hannes Alfven.

As primitive as the plasma theory was then, the concept was revolutionary. The promise that it held, for an infinite energy-rich future for humanity, would have rendered the private ownership of the entropic energy resources in the world, obsolete, such as the various fuels. One of the pillars of empire would have fallen by the wayside if the anti-entropic energy concept had been further developed to the stage of implementation. The private energy resources are presently deployed in the massive looting of society by the oligarchic system of empire. And so, Alfven's work, perhaps without him being aware of it, had threatened to pull the rug out from under the system of empire.

# Big Bang theory to prevent the collapse of empire

It appears that the Big Bang theory was hastily invented and massively promoted, to prevent the collapse of empire by the development of society from within, which would normally have occurred.

# Empire projected its death-model onto the universe

The chokehold on society begins with keeping its center empty, its science. It is said that a great debate had erupted in the 1920s in the halls of empire, between H. G. Wells and Bertrand Russell, over how the danger to empire, of the free development of science, can be prevented. It appears that the promotion of the idea of entropy, projected as the nature of the universe, was the chosen path. With it, the system of empire projected its own inherent death-model onto the universe as if no other model does exist or is possible.

# Empire is an empty hole that drains the world

Empire is an empty hole that drains the resources of the world into its empty center, whereby the world becomes darker, and darker, until civilization disintegrates. At this point empire dies with it. Every aspect of empire is entropic in nature by the platform of the universal thievery on which it operates. Not a single empire in the history of the world has not collapsed itself with its own entropy.

# The current world-empire is no exception

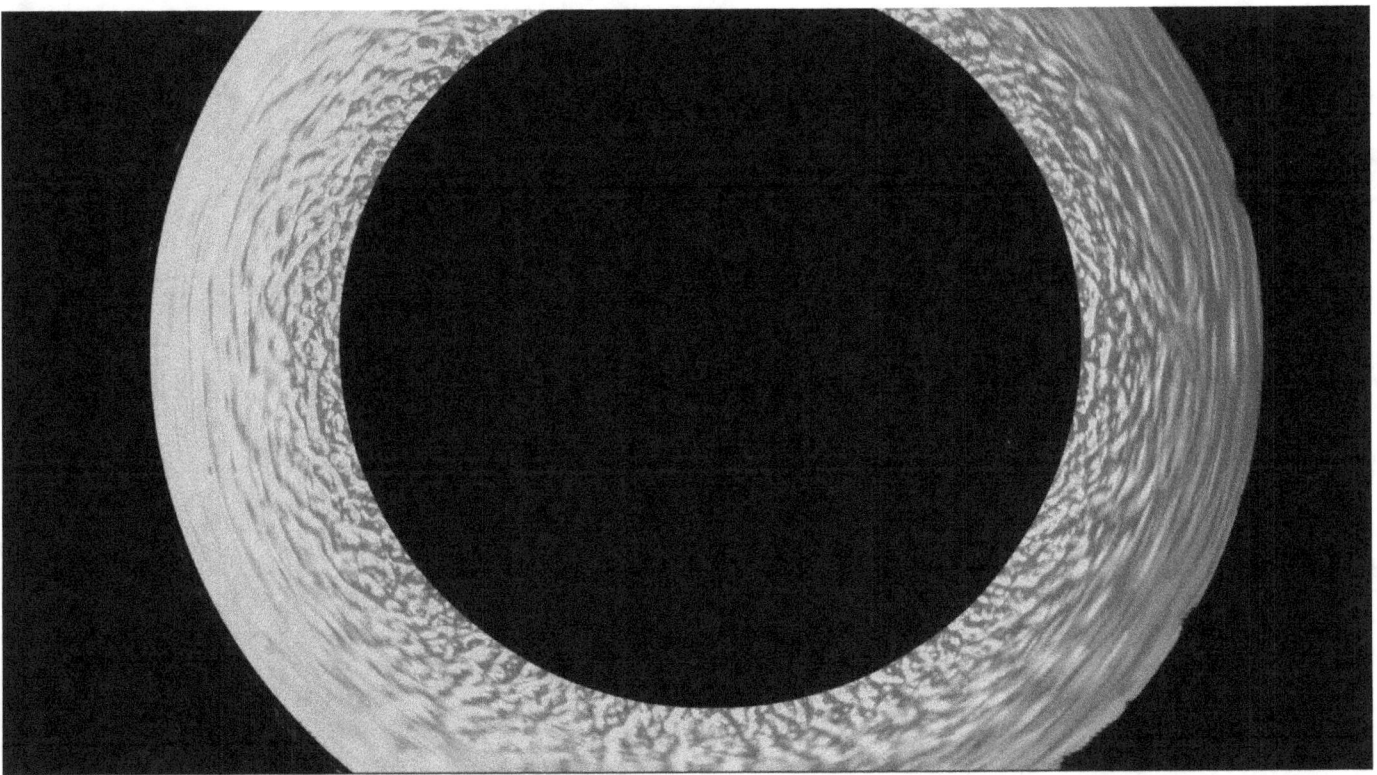

The current world-empire is no exception to the inherent self-collapse of an entropic system. It remains standing in the world as but a ring of smoke with nothing of substance at its center.

# Empire is worse than just being empty

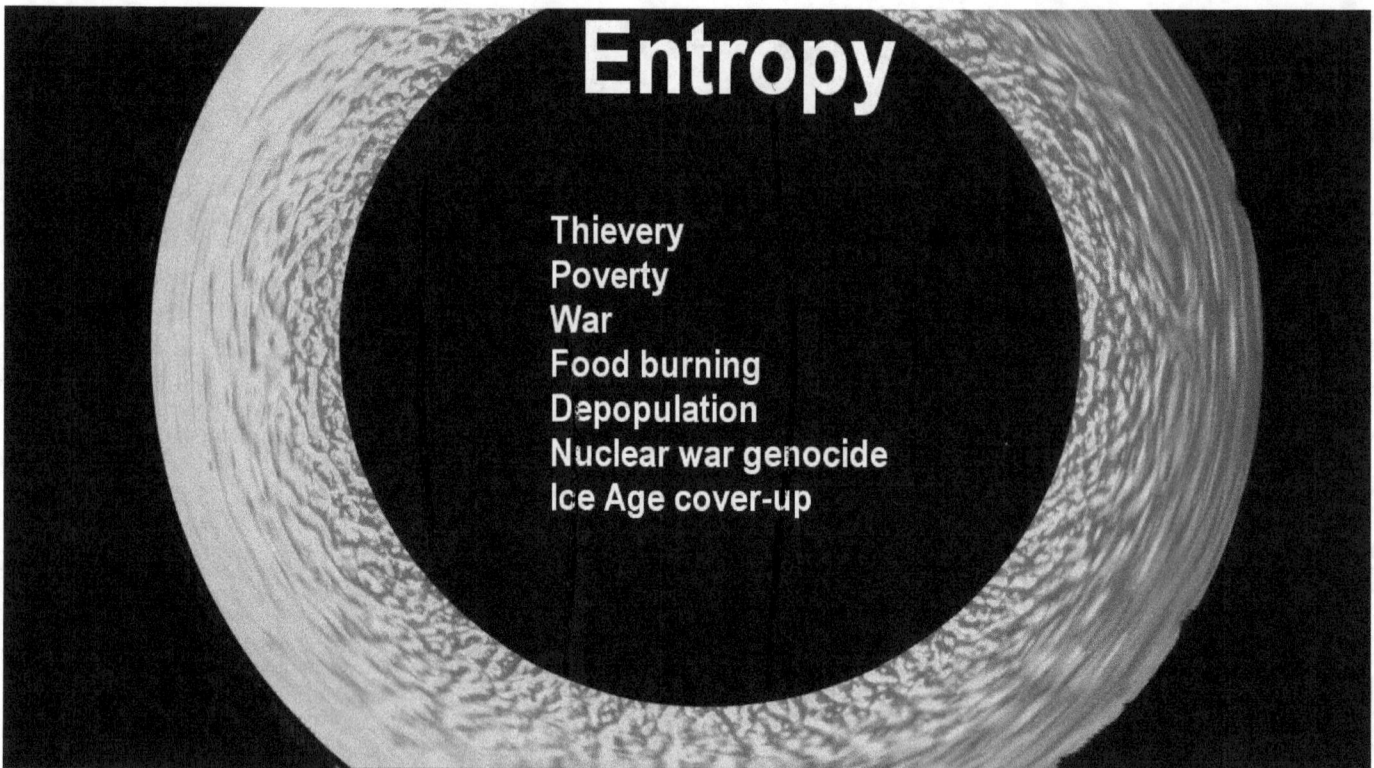

Actually, today's system of empire is worse than just being empty. It is a complex of thievery, poverty, war, food burning, depopulation, nuclear war and terrorist threats. And worst of all its fascist nature blocks the recognition of the near Ice Age that is now before us.

# The Big Bang concept is choking science

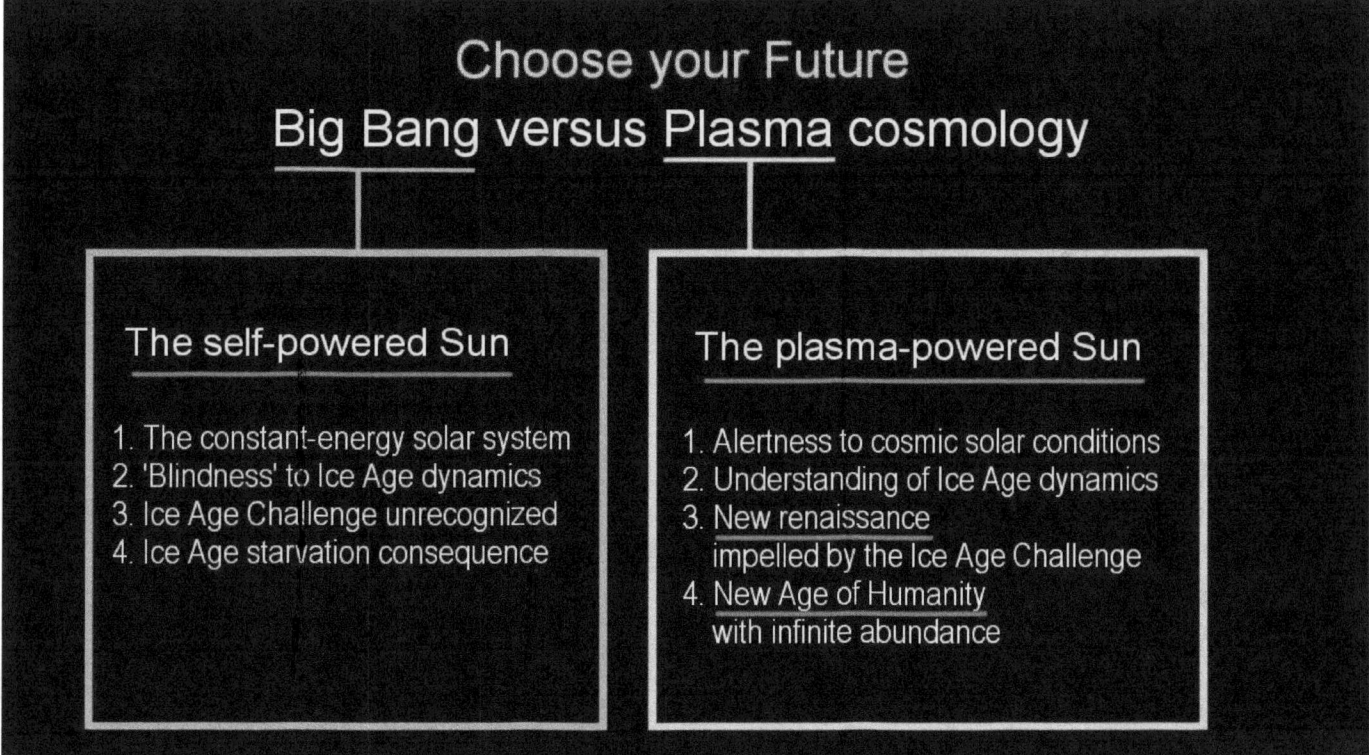

The ultimate form of entropy, built into the Big Bang concept, is its choking effect on science that prevents the recognition of the near Ice Age and its consequences. The extreme lack of scientific recognition prevents the building of the needed infrastructures in the tropics for the future existence of humanity. Under the Big Bang choke-hold on science, 99% of humanity stands poised to starve to death when the Ice Age begins anew, potentially in the 2050s. Only the honest recognition of the science of the plasma cosmology can prevent the consequence that few will survive.

The entropic Big Bang explosion, that theorizes a 'beginning' before which there was nothing, thereby inspires an 'end' for everything. It is a cleverly contrived small-minded concept that appears to have been intentionally staged to induce small-minded, empty thinking, into cosmology as a means for keeping a lid on the progressive development of the plasma-universe concept that the system of empire cannot survive, which as of old will take humanity to the grave with it if humanity remains latched onto this empty system. That's the choice before us.

## Anti-Entropy in Civilization

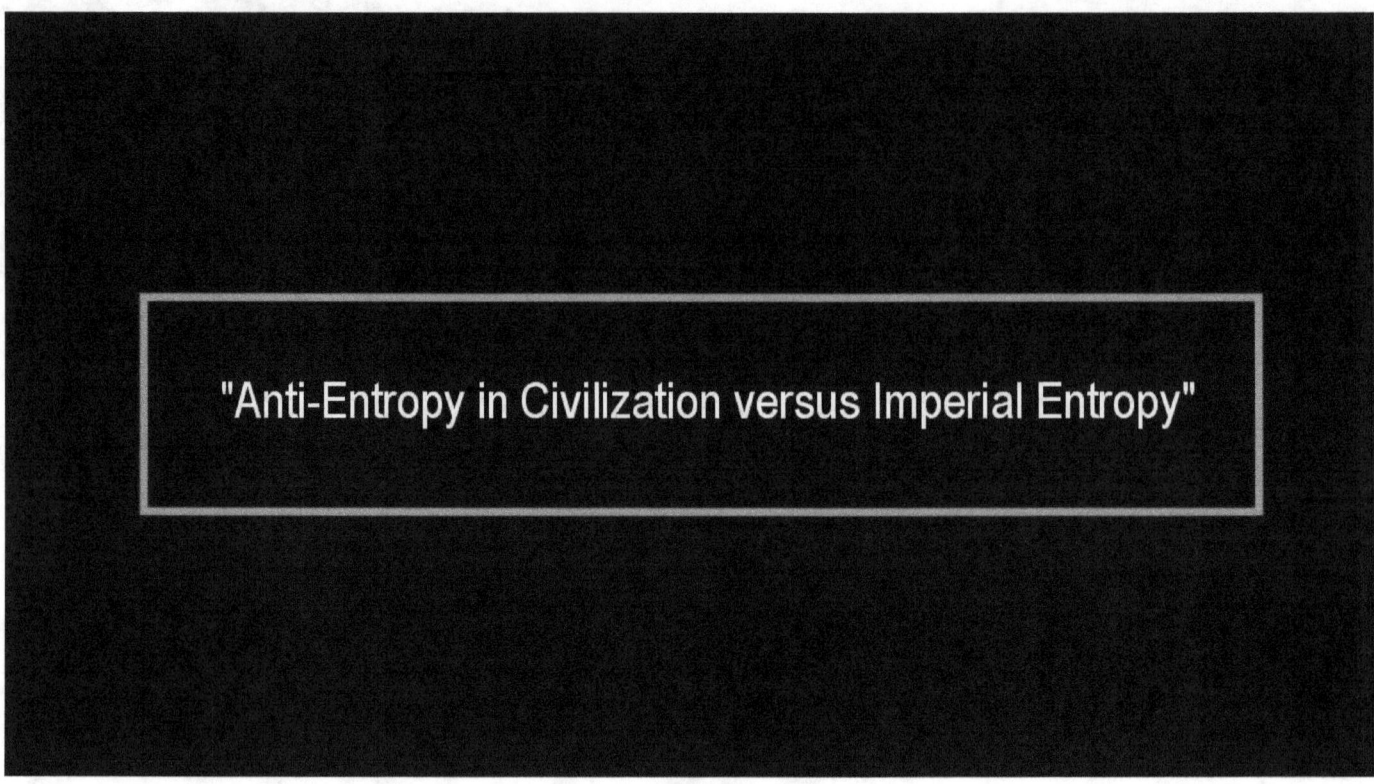

"Anti-Entropy in Civilization versus Imperial Entropy"

# The human bang, the real big bang

Should we not rather choose the anti-entropic 'heaven' of grand discoveries that spark a revolution built onto the creative and productive power in society, expanding into all directions?

That's where the human bang, is located, the real big bang. This is the choice before us, which the Ice Age Challenge should inspire. This choice would spark an explosion of ideas and subsequent creations. For example, with basalt as the feedstock for a new industrial revolution, and it being shaped in automated high-temperature large scale industrial processes, we have the power at hand to create the needed 6000 new cities with ease, for a million people each, to meet the Ice Age Challenge. And those would be free cities, with free housing for all. Also, the 6000 new cities could all be built in time to enable the relocation of the northern nations prior to their territories becoming uninhabitable. All of this becomes possible when we choose to step away from the entropy of the dark hole of empire.

## We can get to this stage of freedom

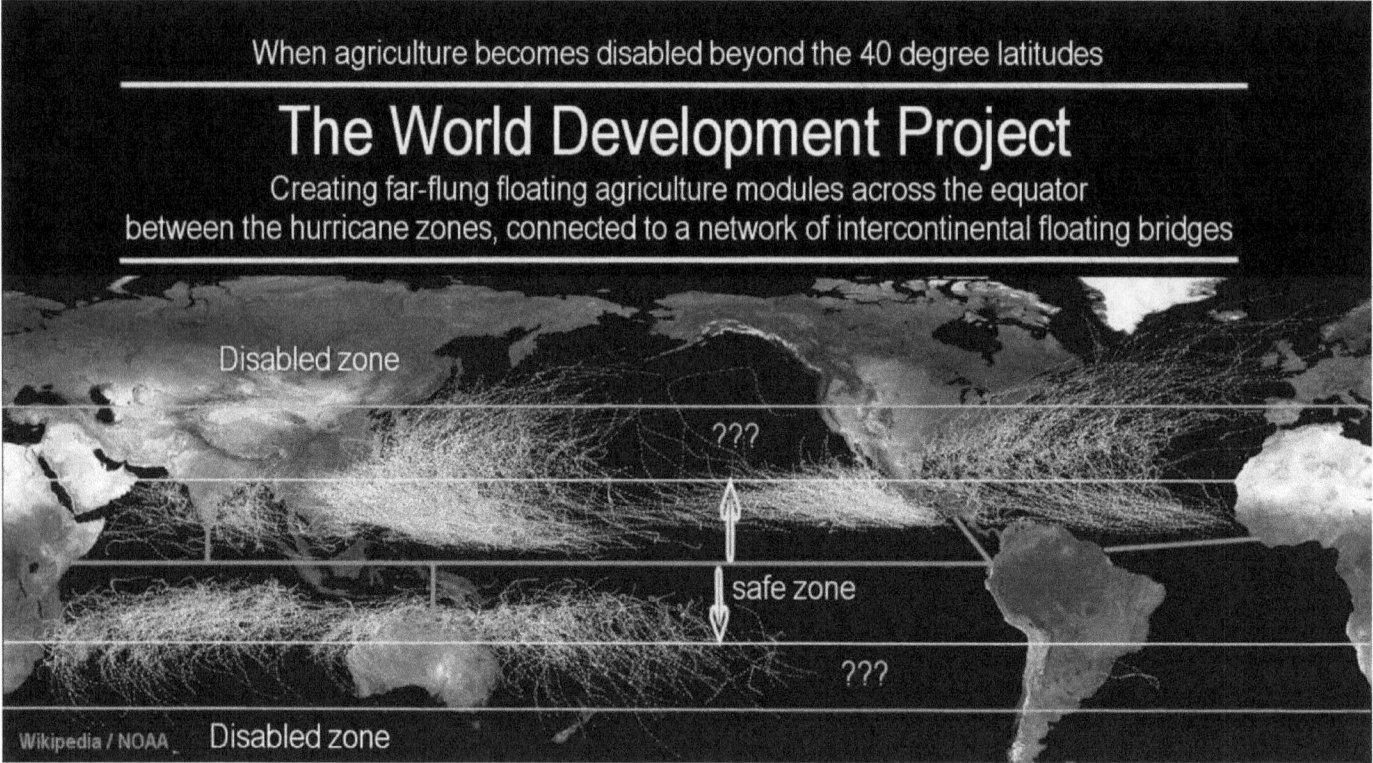

We can get to this stage of freedom relatively easily, by not allowing ourselves to be enslaved to the imperial platform. Once this decision has been made, society will discover its creative power and also relocate all of its endangered agriculture into the tropics before the Ice Age begins, placing it afloat on floating modules made of basalt, interspersed with floating cities that offer universal free housing, and with vast new industries attached that are needed for meeting the human needs.

All of this is physically possible right now, with the already developed technologies and readily available energy resources, for the utilization of the near infinite volumes of the highest-grade materials that are sitting process-ready, unused, on the ground. The vast potential for this type of revolution in human living, which is already critically needed in the modern world, is not being blocked by any physical limits, but is blocked by the 'devil' of entropy in the kingdom of empire that, in its numerous ways, rules the world.

# In the shadow of blocking of humanity

In the shadow of this blocking of humanity, civilization is fast breaking down. It is choked by the composite 'devil' of 'empire and entropy' that operates as a single package. The result is the collapse of civilization towards nuclear war, while humanity's vast human potential remains dormant, unrealized, and its future is squandered and is becoming potentially lost.

This tragedy is not what our choice should continue to be, which leads to extinction in several ways.

## Our present stage is precarious

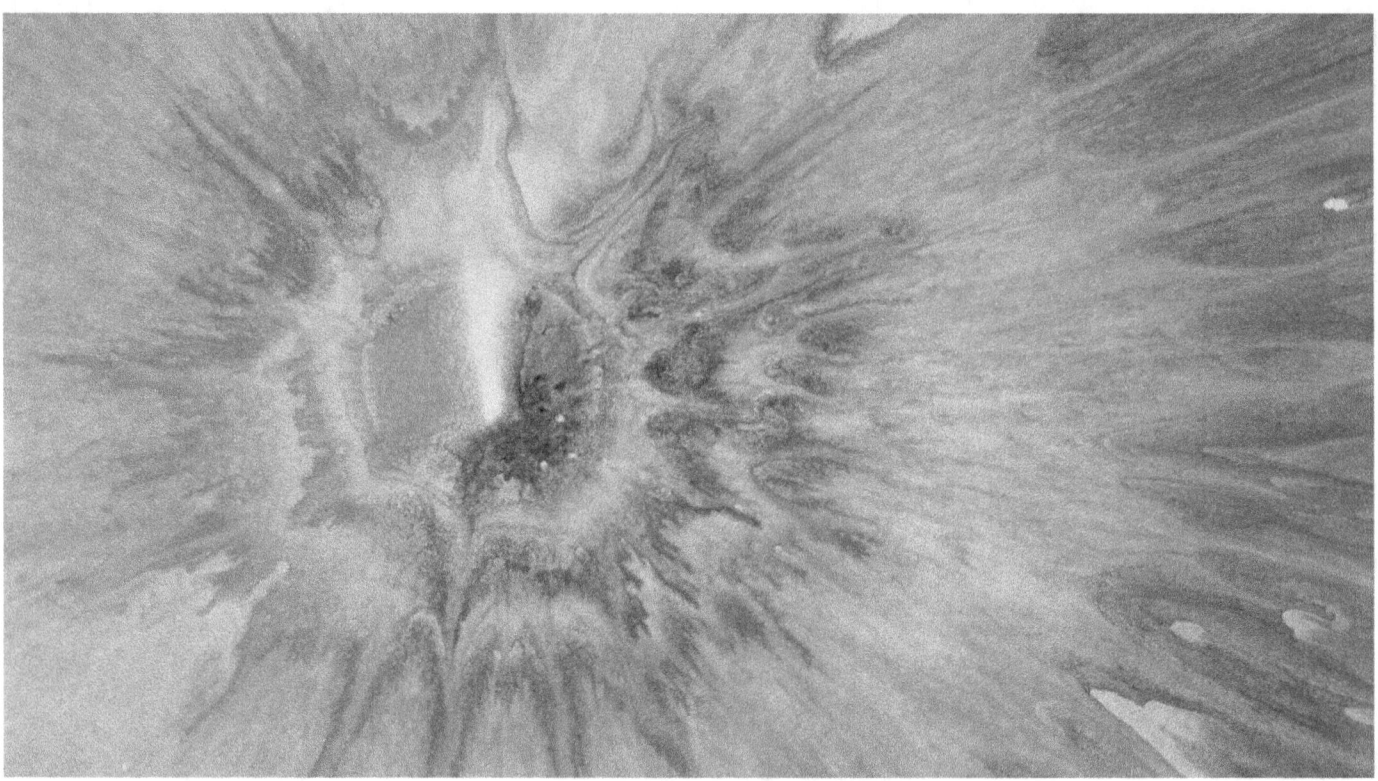

Our present stage is precarious, indeed, but it is so only, because the spark in the heart has not yet exploded into the great world-enriching fire that the term, humanity, should inspire. When this spark in the heart will awaken and live up to its inherent promise and light the fire in the heart, our civilization will have a wonderful bright and colorful center that is rich, and beautiful, and substantial, and anti-entropic.

# Running away from anti-entropic economics

Society is presently running away from anti-entropic economics, even while it is the only platform that actually exists for building a civilization on. However, society cannot evade the consequences that ensue when it is running away from this only platform.

## Not a platform for civilization

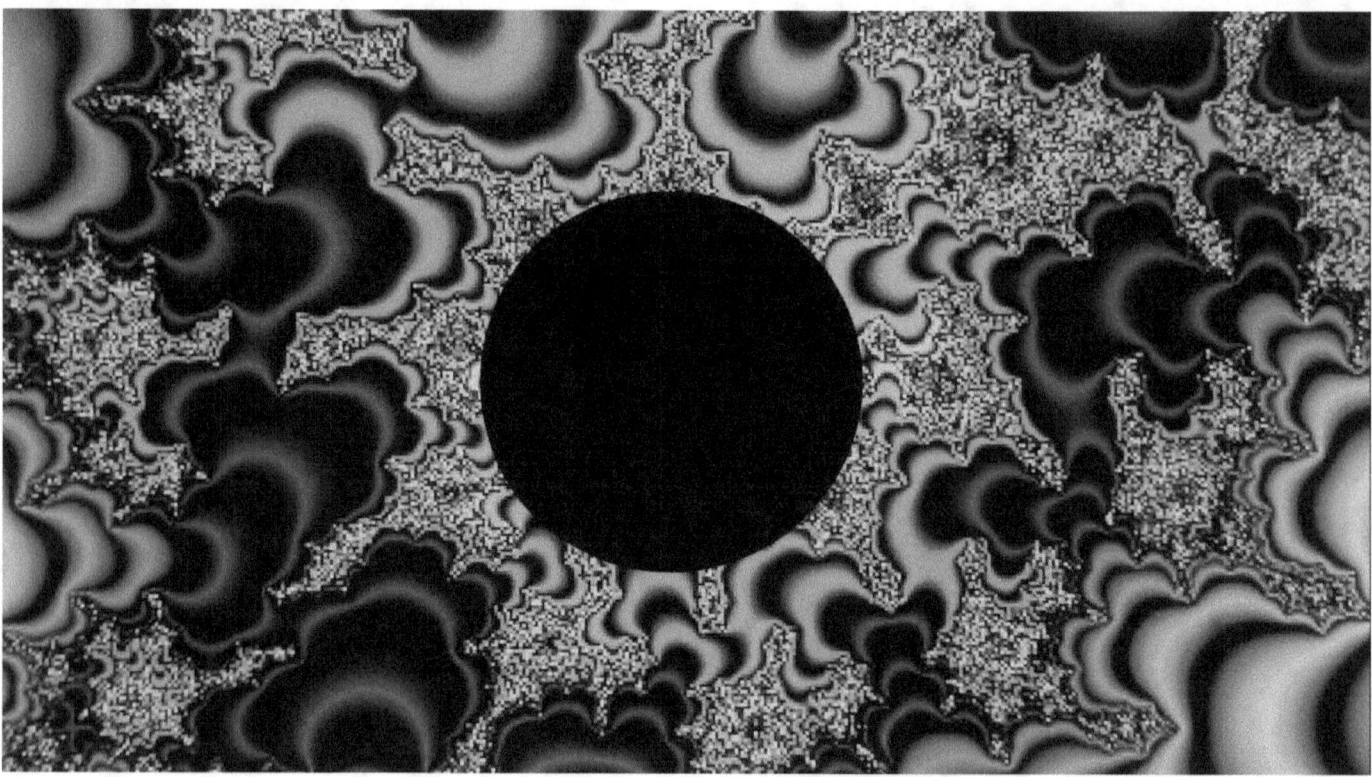

The entropic platform is not a platform for civilization. It is the platform invented by empire in support of its looting operations. Nor is the entropic platform, a platform for economics. It is not designed for building anything on it. It is designed for diminishing everything, for tearing everything to the ground. It is the platform for the money bags, that's required for sealing and for financial derivatives that have the effect of a nuclear bomb in economics. The entropic platform is a power tool for looting that wrecks everything, in contrast with the anti-entropic platform that builds everything.

# Two Opposing Platforms in Economics

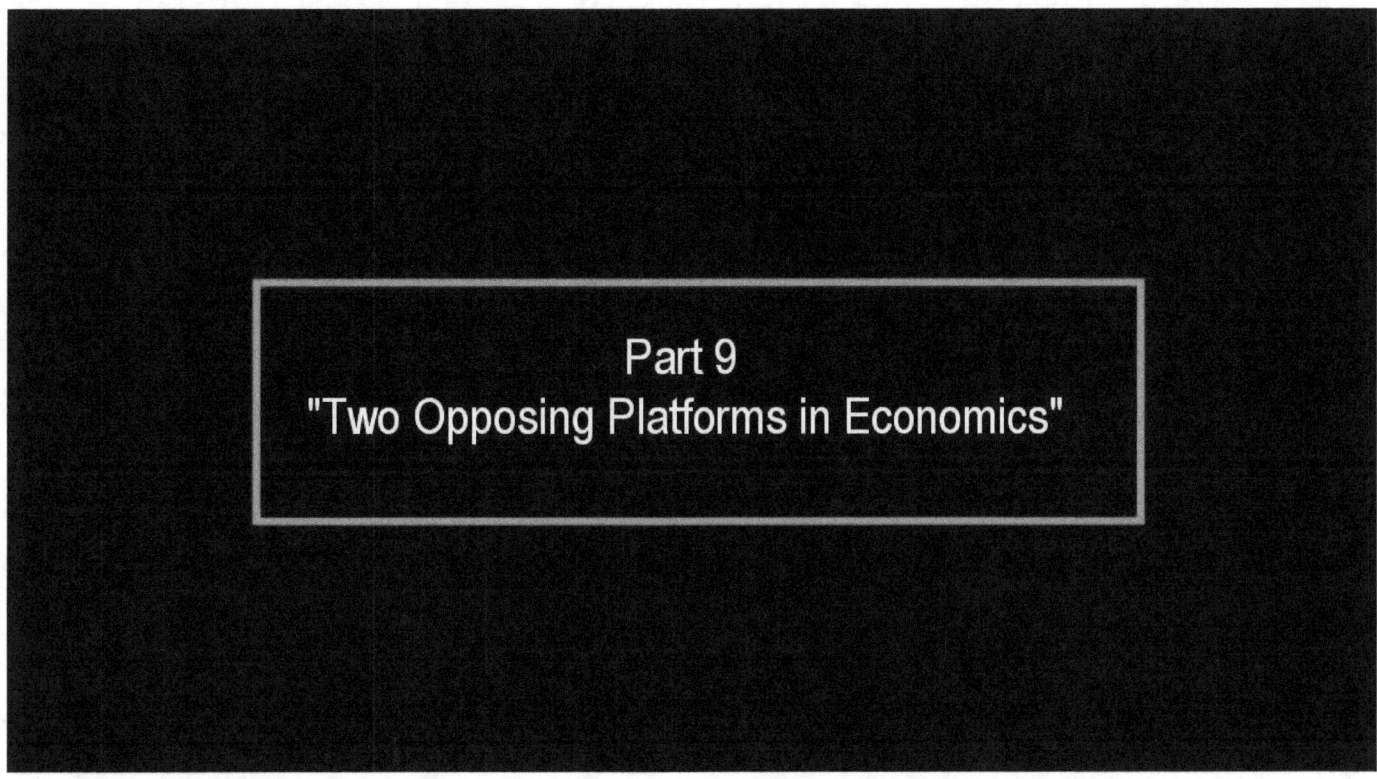

"Two Opposing Platforms in Economics"

# The anti-entropic platform

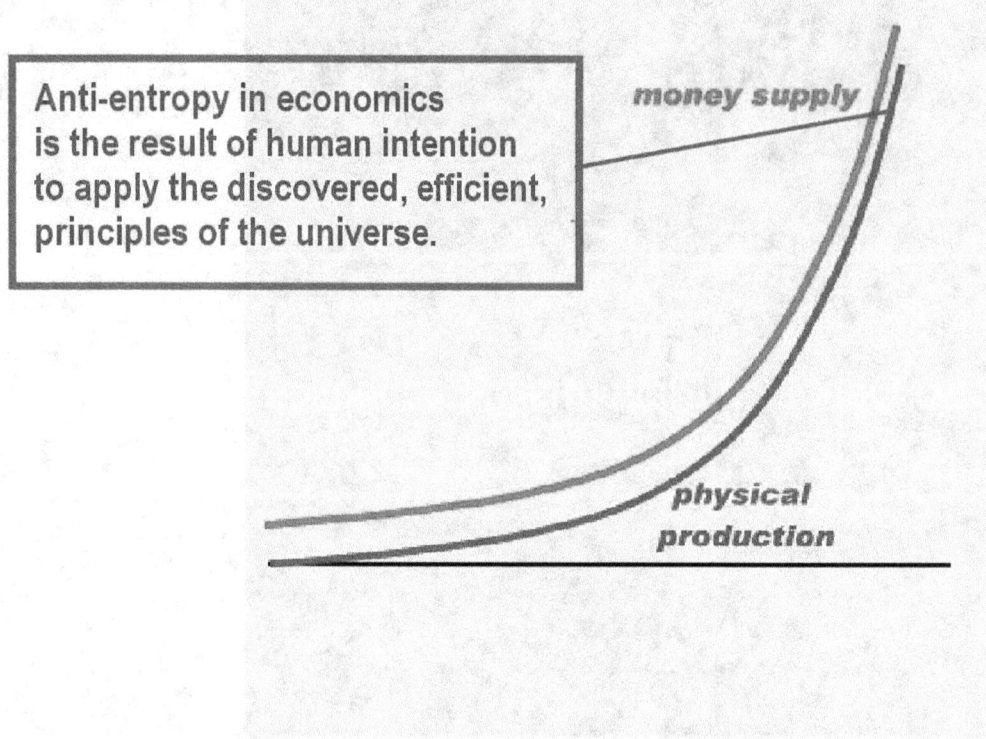

Let me illustrate what an anti-entropic platform looks like, in principle, that is for building a civilization.

The anti-entropic platform is a platform for creating value for human living, by applied human ingenuity and productivity. The anti-entropy, the expansion of value in an economy, is in this case the product of intention. The intention is, to apply the discovered principle of the nature of the universe, to the building of a civilization. The principle that the universe is built on is evidently the most efficient principle. On the resulting platform of intention, society provides to itself whatever financial credits, material, and human resources are required to fulfill its specific needs. Nothing is borrowed here. Nothing is stolen. Every need is met by creative and productive processes. The result gives money its value.

Note, that the factor of intention is of critical importance here.

Anti-entropy in economics is the result of human intention to apply the discovered, efficient, principles of the universe, to human living.

# A bridge across the tropics

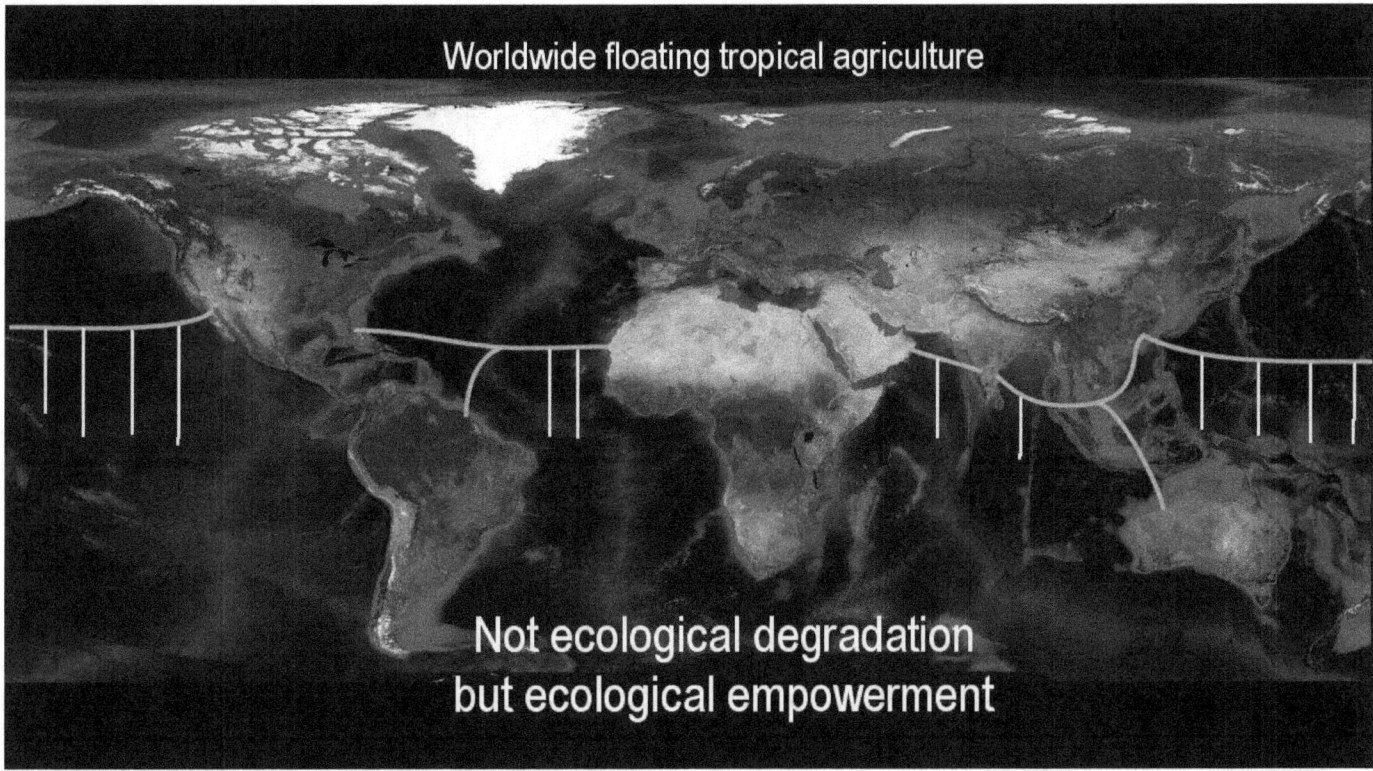

For example, if a bridge across the tropics, with floating agriculture attached to it, is required for society to have a rich and secure future, then a national bank will be founded that provides the financial credits to get the job done. When the job is done, the obligation by society to itself is extinguished. The credit that a national bank would utter for such a project, would not be regarded as a debt to be repaid, because the value created for society, by the completed infrastructure, has at this point already extinguished the obligation by society to itself. That's revolutionary, isn't it?

# Credit to get an enriching job done

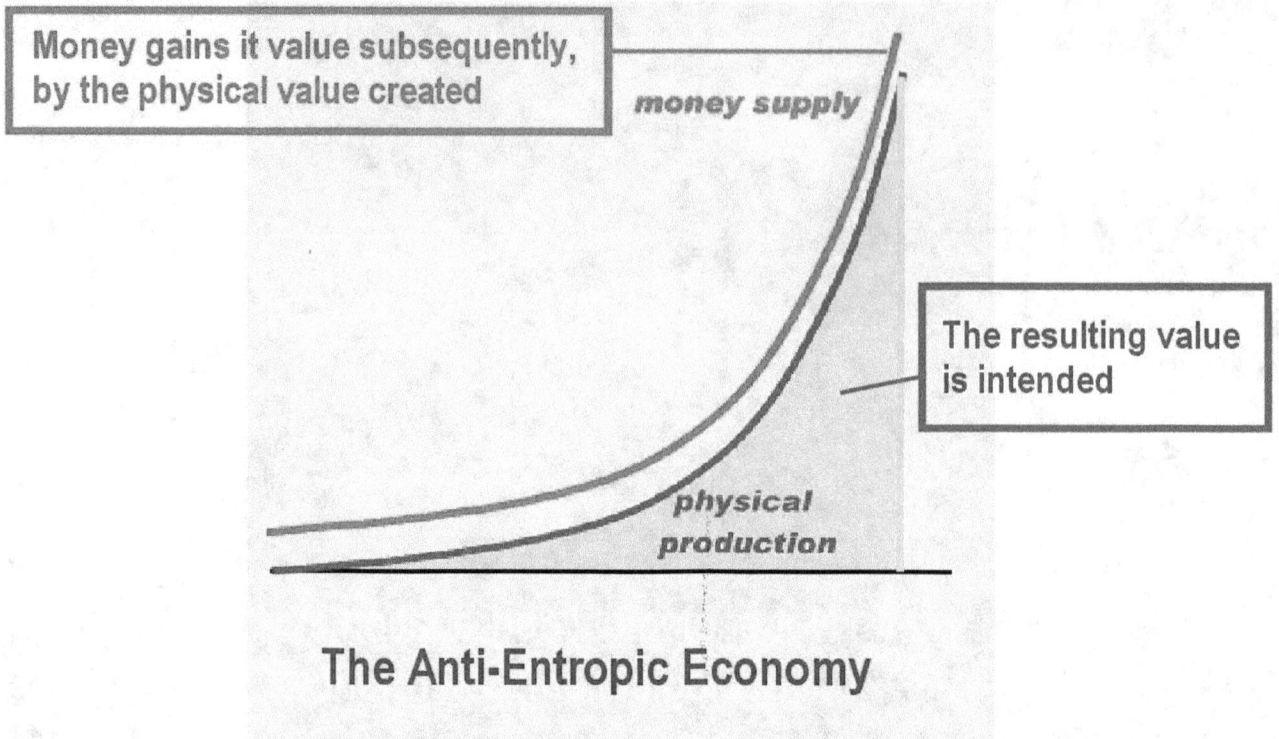

If money is created as credit to get an enriching job done. The resulting value is the object that is intended. Money gains its value subsequently from that. It gains its value from the physical value created that reflects itself in a richer life in society. The richer life is represented by the shaded area. Which is the intended value.

The process may not even involve the creation of money as credit. Other forms of compensation may be applied to get a job done. The focus is on the intention for physical value to be produced for society, whereby its living becomes richer.

The anti-entropy in this process is not located in money itself at any point, but is located in the physical value created by the human creative and productive process. The entire process is driven by the intention to meet the human need, and secondarily by the intention to provide the needed credit, or promise, or commitment to get the job done.

# Economics in civilization

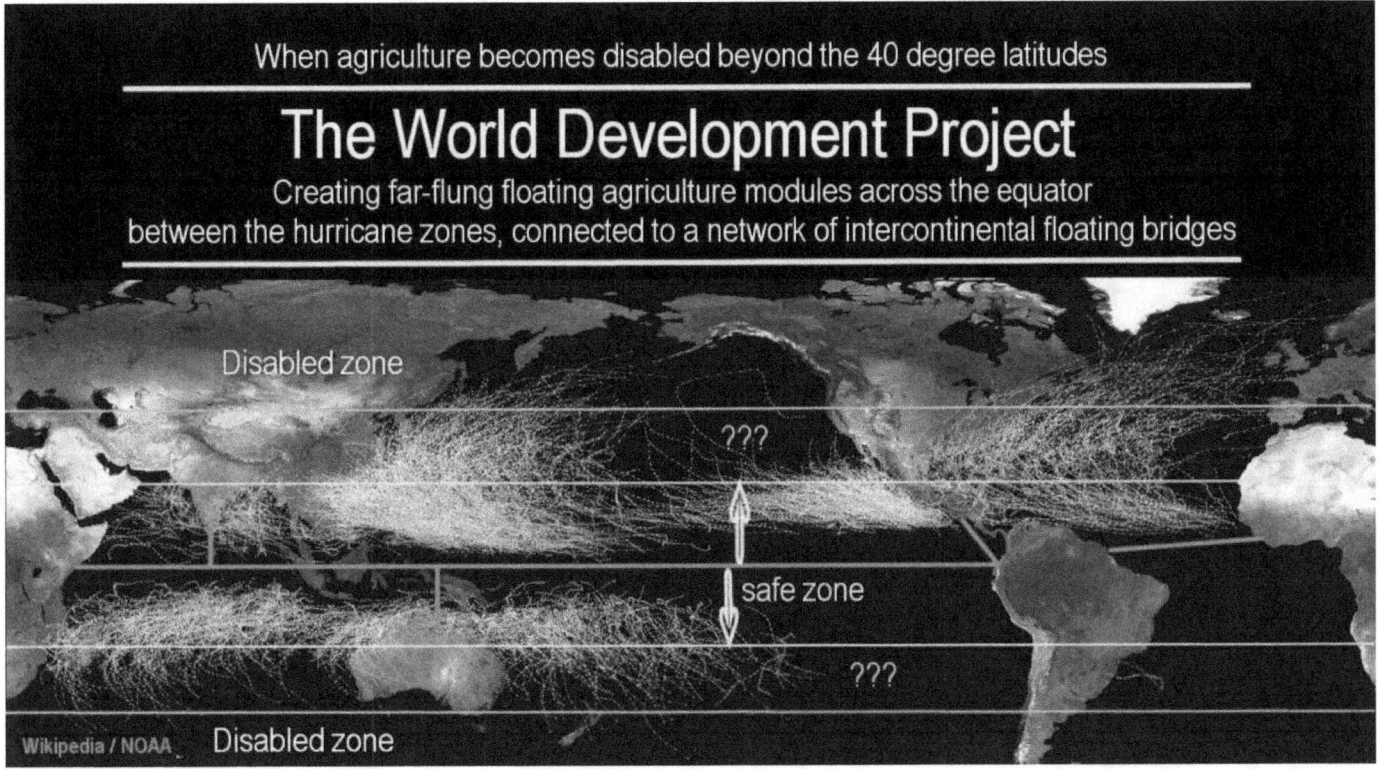

This type of economics in civilization will never be focused on what it will cost to create a richer world. It will be focused on the value that can be created for civilization. The key question will then become, what else can we build for us to make our world grander, our living richer, and our existence more secure?

And so it will be, that increasingly more and more will be built, for ever-greater value being created for society.

On this anti-entropic platform that increases to productive power, the Ice Age Challenge will be met with ease, and may even become a sideline issue. The financial credits that may be uttered to get the job done, will of course never be repaid, as the debt has become extinguished when the job is done, that is when the objective has been fulfilled.

The revolutionary principle - that the debt that society owes to itself to create world-enriching infrastructures, will be extinguished when the infrastructures are completed - totally eliminates the private estate monetarist platform. No money bag will ever lend a penny on the platform that nothing will be paid back in money, nor any interest that it seeks for profit.

# The money-bag system

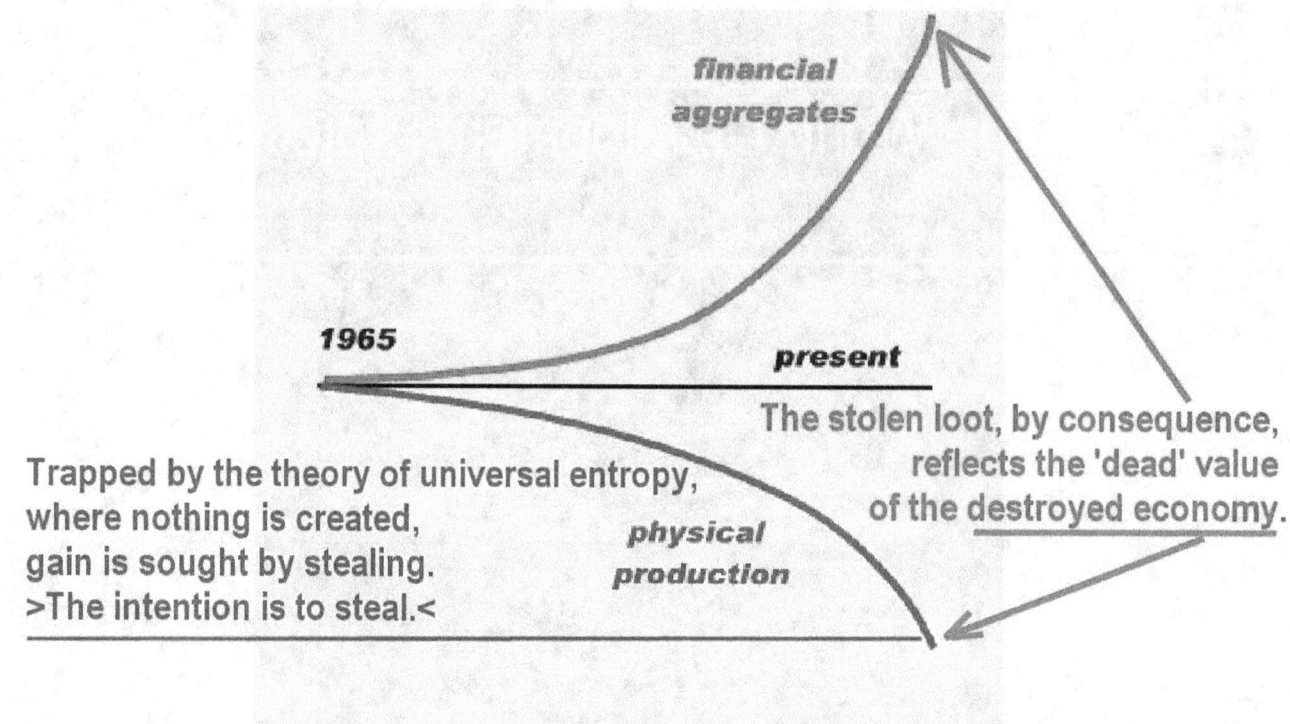

The money-bag system has a different intention standing behind it, than to seek the development of humanity. Its intention is to swell its bags, instead of producing value for society. The intention is to steal, instead of to produce. The intention renders the money bag system entropic in nature. Trapped by the theory of universal entropy, where nothing is created, gain is sought by stealing, and so on. This characteristic applies to almost all forms of monetarism, from the stock markets wagers, to financial derivatives gambling, to international currencies speculation, etc.. All of these markets are huge, where nothing is produced, though enormous profits are drawn.

When profit is claimed where nothing is produced in value, the profit is essentially stolen from the physical living of society. However, the processes of stealing, which diminishes the productive power of society, as the graph shows, also diminishes the value of the stolen loot, regardless of its volume, as it stands as a claim against the product of the collapsing economy. The stolen loot, by consequence, reflects the 'dead' value of the destroyed economy. The bottom line is, that the widely held belief in the false theory of entropy, has unavoidable entropic consequences. And those are tragically real.

That's what is illustrated here. The illustration presents in principle the entropy inherent in the oligarchic monetarist system. Every empire is tied to this collapse function by its very nature. The illustration was developed at around 1995 by the world-renowned American economist Lyndon LaRouche for a presentation at the Vatican. LaRouche illustrated in principle 20 years ago the dynamics that the western world is experiencing today. Our tragic experience is the result of the general belief in a false theory that has become accepted and assumed to be beneficial, while the opposite is the case.

## Empire is doomed by its own premise

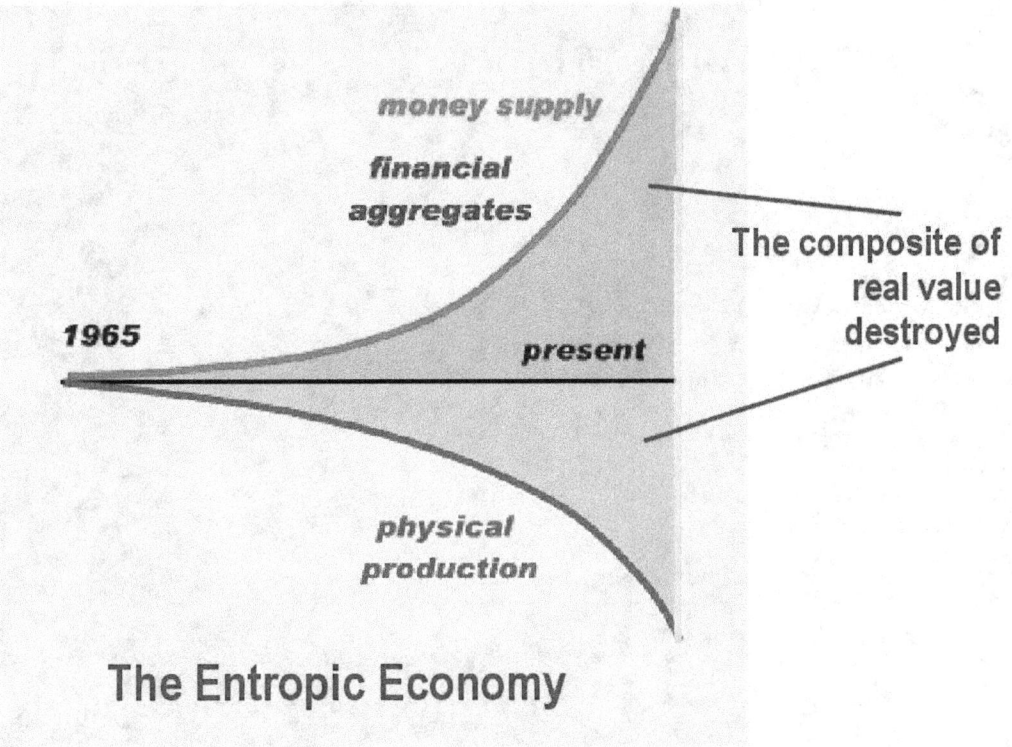

The shaded area between the two curves, represents the destroyed economic value by the collapse process. The destruction of value is typical for an oligarchic, imperial, monetarist system. No other result is possible in an entropic system that the system of empire is, where nothing is created and everything is doomed to diminish to zero.

Every system of empire has doomed itself on this platform, for the simple reason that an entropic system cannot be cured, or be maintained.

The system of empire is doomed by its own premise. It is also doomed for the related fact that no money bag will ever regard the richer living in society as the proper fulfillment of the terms for credit uttered. The money-bag system demands the pound of flesh it claims for profit. In a living world, this claim will never be paid. For this reason, the bag-money system will simply vanish off the face of the earth in the near future, and be replaced with an anti-entropic love-coin system that produces value for society.

The key difference between the two types of system, the anti-entropic system presented earlier, and the entropic system presented here, is a difference in perceived theories that guides the intention. Every system of empire, regardless of its form, is built on the false theory of universal entropy that invites stealing, which thereby collapses itself accordingly. Nothing can save this system that was functionally dead from the outset. Neither war nor depopulation can save it from its built-in fate. Its doom lays in itself by the entropy that it has built on. Society, however, has the power to unlatch itself from this doom, and hasten it.

# The consequences are world-destructive

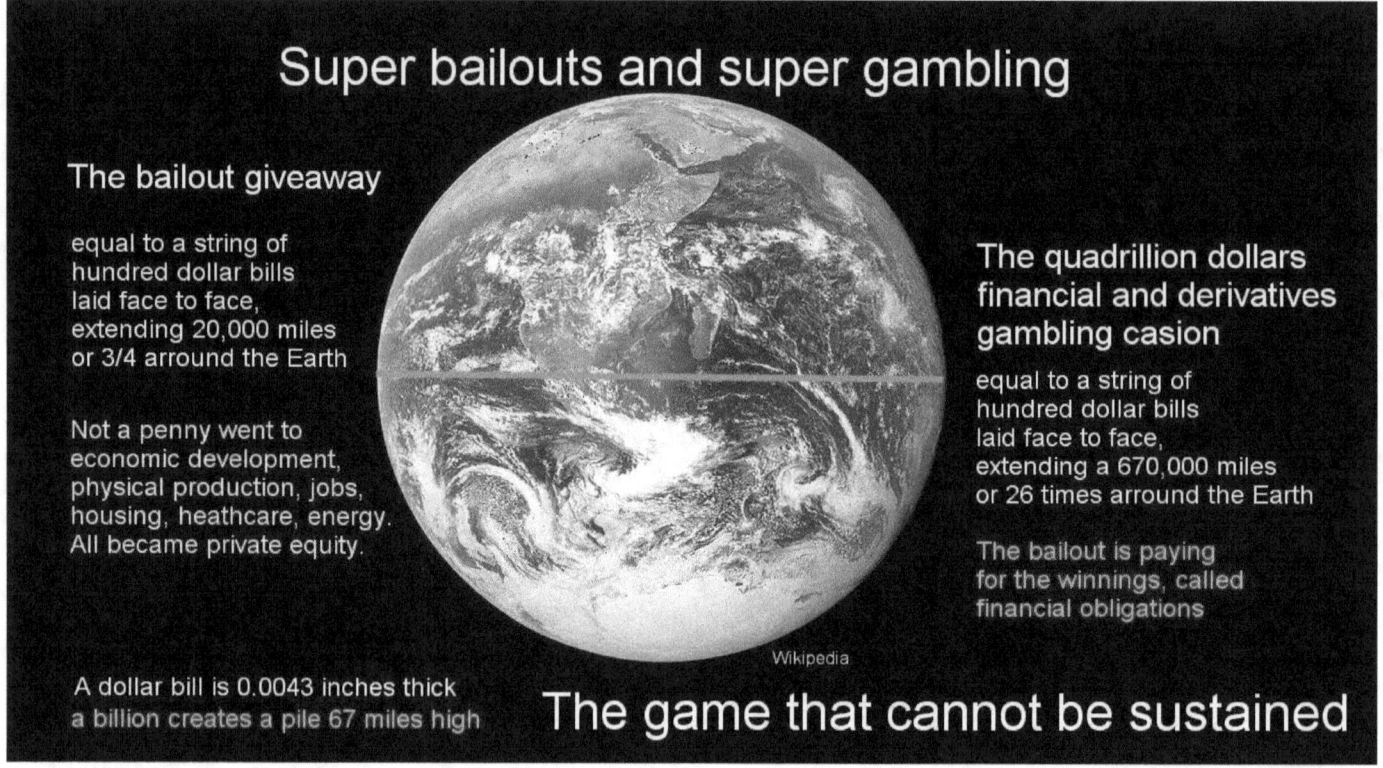

For example, in a gambling arena where 1 to 2 quadrillion dollars are riding the dice, the consequences are world-destructive. For civilization to survive, the entire entropic gambling system needs to be scrapped. And this means, scrapping the system of empire in its entirety, including every law that stands in support of it, without fail.

# The anti-entropic system of economics

The anti-entropic system of economics, which reflects the natural creative principles of the universe and man, offers life instead of doom. In its context, science and cultural development are promoted as elements of the anti-entropic system, because science and culture are the key driver for increasing the productive power of society. Real science is anti-entropic. It unfolds the quality of humanity that is without limits.

It was asked in ancient times, how can one know what God is? The questioner was asked to look at the tip of his finger, who does it point to? It points to me. There is your answer, the questioner was told. God is reflected in man. You know no more of God than you know of yourself. So don't belittle God in reaching for the maximum of good, and don't belittle your power in achieving that maximum of good. Here, science begins. With science as a part of you, you can begin to see the universe as it is. And what you see has no limits. The universe is self-powered, expanding, unfolding, and developing itself. And that's what you see when you look at the point of your finger. You see your humanity as the image of God. You see yourself as a part of the process in which you are empowered, and are expanding your proof in the world that you have lived and have developed beyond your wildest dreams.

Free high-quality housing, health care, and education, are factors in this process of unhindered advance in science and culture. Free high-quality housing, health care, and education, are not seen as liabilities in an anti-entropic system, but are seen as opportunities for advancing the creative power of the human being, and thereby of society itself as a whole, which should be promoted and be developed to the highest degree, to maximise human living. In an anti-entropic economy, slum living, poverty, unemployment, diseducation, and other forms of deprivation, stand as economic crimes against society, because they tally up to the deadliest waste possible, the waste of the most precious asset that a society has, which is itself.

The question may be asked here, "Will we get to this point where entropy in economics, and with it, empire, are history?" Of course we will get there, because nothing less will enable humanity to meet the Ice Age Challenge in the few years that we have still remaining till the next Ice Age starts.

## We are the supreme being on Earth

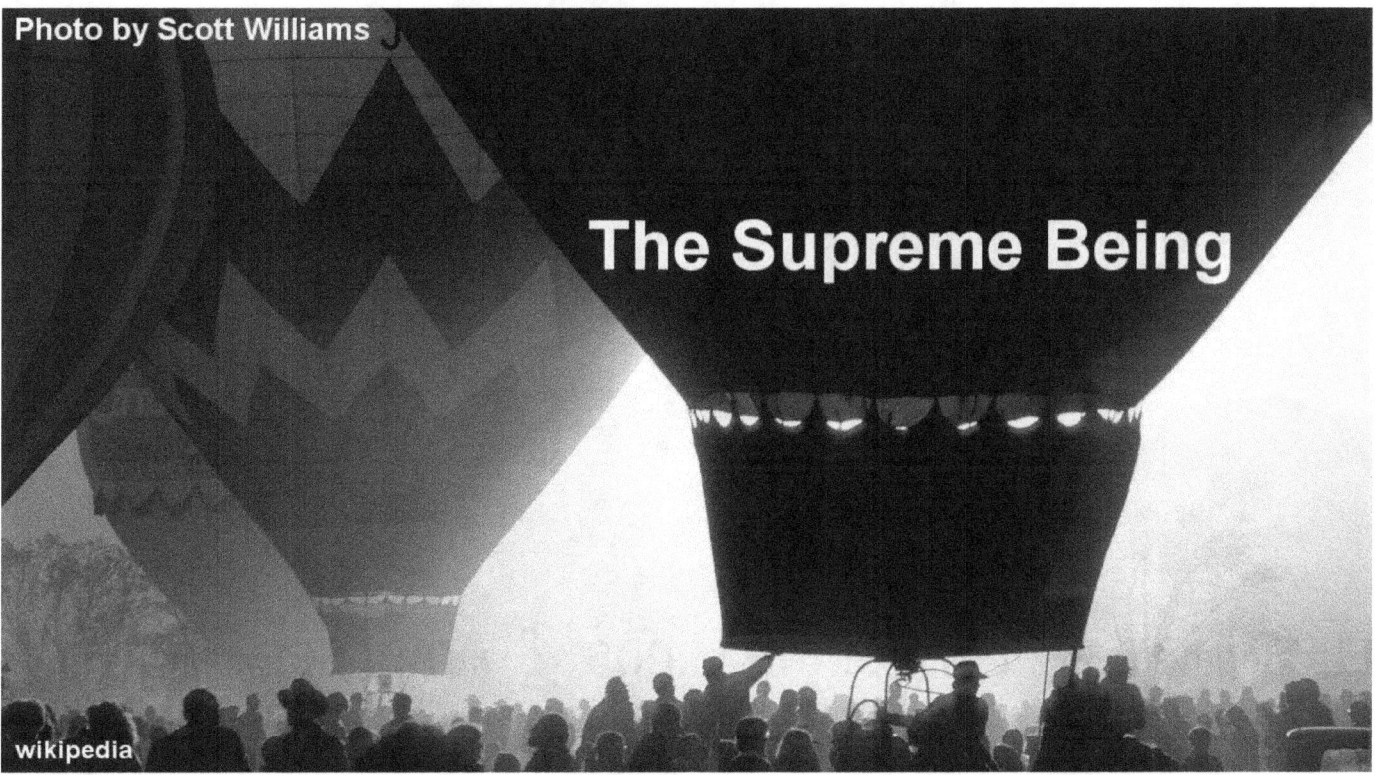

We are the supreme being on Earth, second to none. If we won't inspire us to recognize our infinite potential, nothing will, and we will remain doomed. Thus, the present realization of our infinite future is not blocked by any physical limits of the universe or limits on the Earth, but is blocked by the smallness at heart that society all-to-often allows itself to become trapped by. In this the poets have been correct, who have said in countless different metaphors that the key to the heavens lies in ourselves. It truly lies there, in every respect.

# Segment 3 - "Lord of the Rings" & The American Paradox

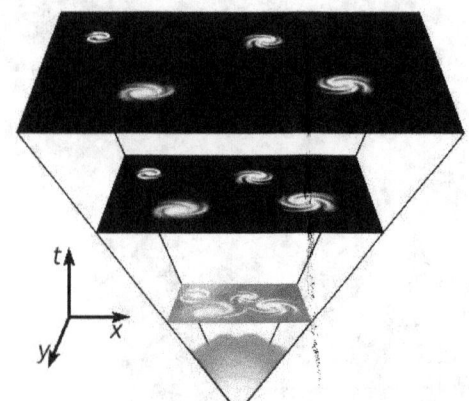

Big Bang Blow-Out

Segment 3 - "Lord of the Rings" & American Paradox

Uphold the platform of civilization without compromise, instructs Tolkien. America compromised with the Glass Steagall law, in 1933, and now suffers the consequences.

Glass Steagall was a compromise. It allowed the empire of monetarist looting to coexist with anti-entropic economics. By the compromise that law was doomed from the outset. A civilization cannot stand on two opposite platforms, the platforms of entropy and anti-entropy. Empire is an entropic system of legal stealing via monetarist looting, such as Wall Street, that tears economies down. America was built on the anti-entropic platform of national credit to uplift the general wealth and welfare of society by its self-development. This platform was compromised by allowing the 'Empire of Wall Street' to continue in 1933.

## Humanity is anti-entropic in nature

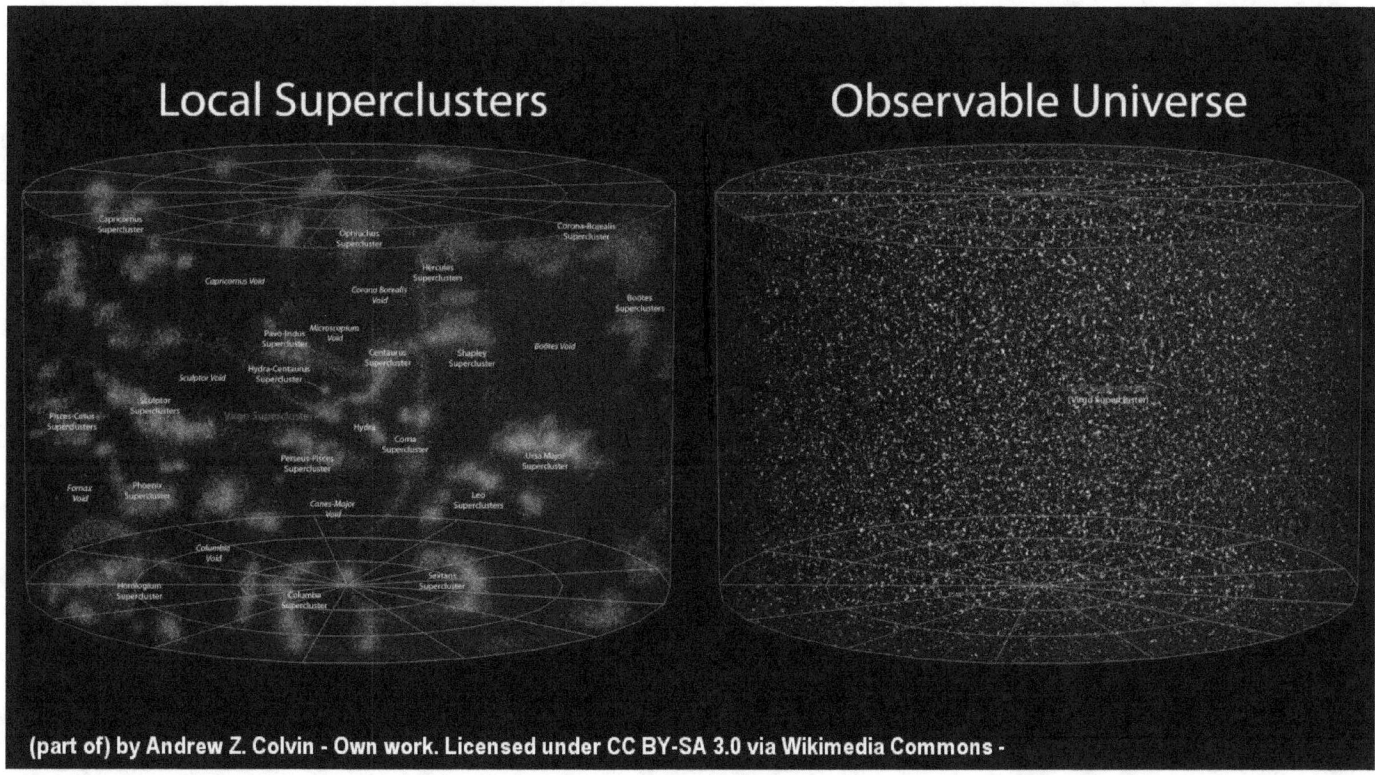

Humanity is anti-entropic in nature and lives in an anti-entropic universe. The numerous forms of real evidence of the anti-entropic nature of the universe, all disprove the Big Bang theory.

## Society's illusions about the universe

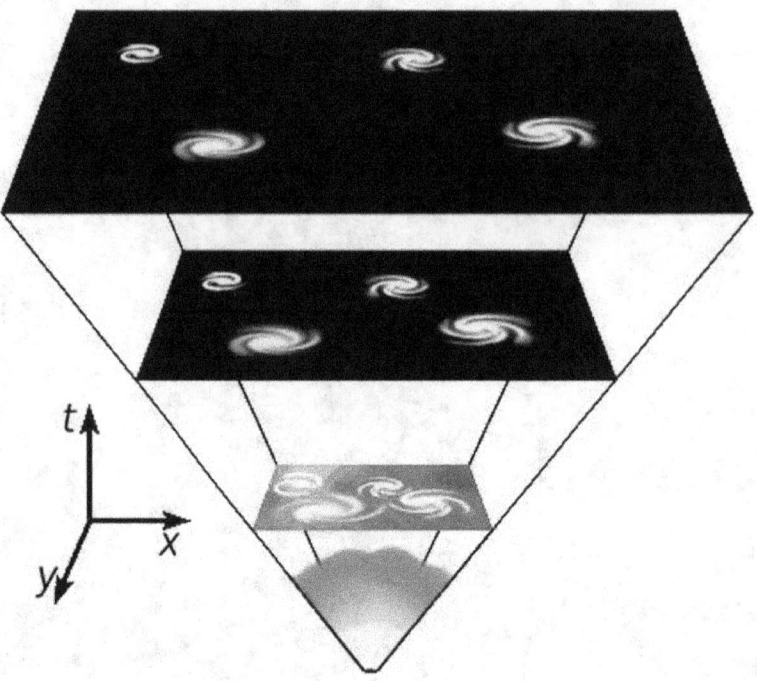

The evidence blows the theory away as a bubble without substance.

Society's illusions about the universe have never changed the universe itself, they have only affected the way the universe was perceived and its principles applied.

# In the Big Bang dream

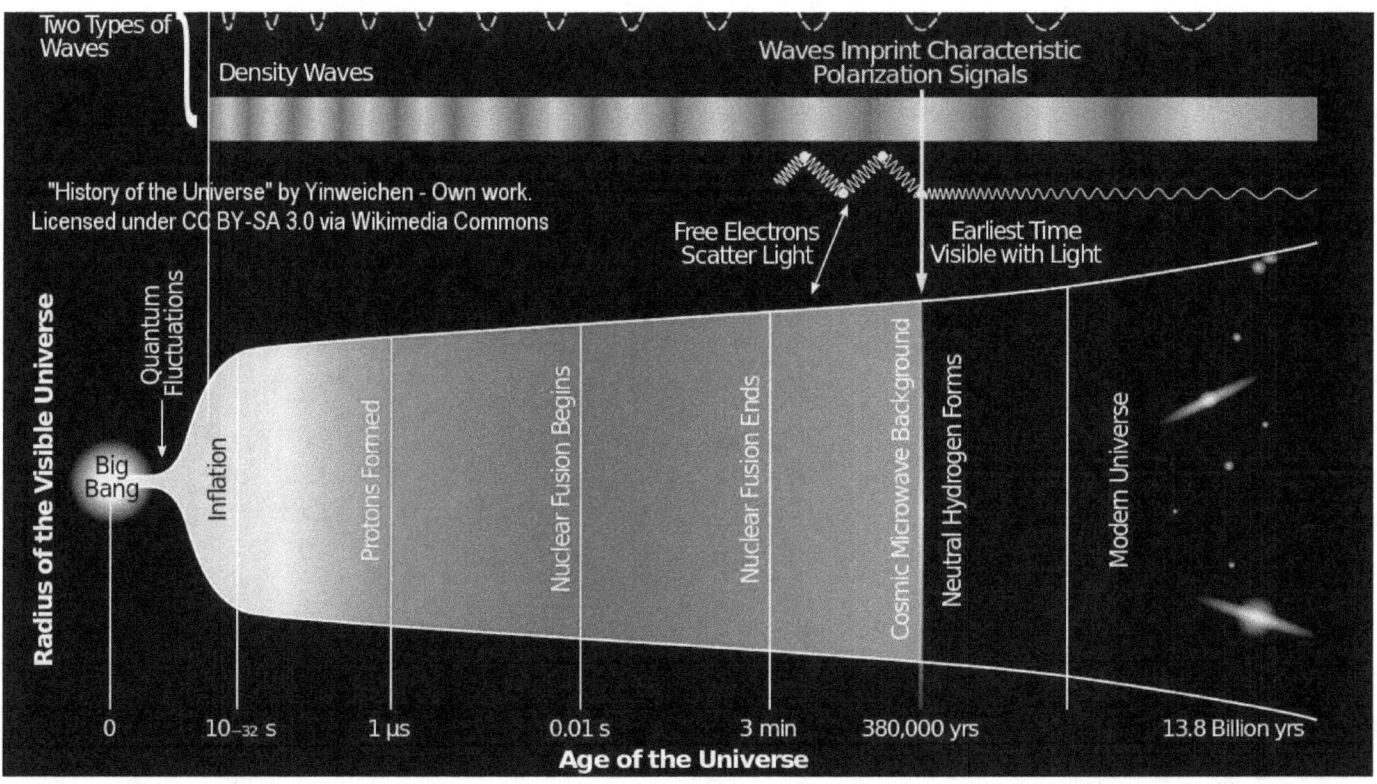

While there is nothing of substance to hold on to, in the Big Bang dream, its forest of confusing illusions is nevertheless, still, rather dense. Consequently society finds it difficult to trace its way out of the forest to the recognition of the real principles of the universe.

## Whether humanity achieves the rich future

Whether humanity achieves the rich future that it has within reach, inspired by advances in science based on real evidence, or whether it dies of starvation in the forest of illusions when the next Ice Age begins in potentially the 2050s, depends on the choice that society clings to in the present.

# The kind of choice that Tolkien places before society

That's the kind of choice that Tolkien places before society in his saga, The Lord of the Rings. He aims to inspire an uncompromised choice.

## Unfortunately our world is darker

Unfortunately our world is darker, than Tolkien's fictional world.

# Our civilization is choked with false theories

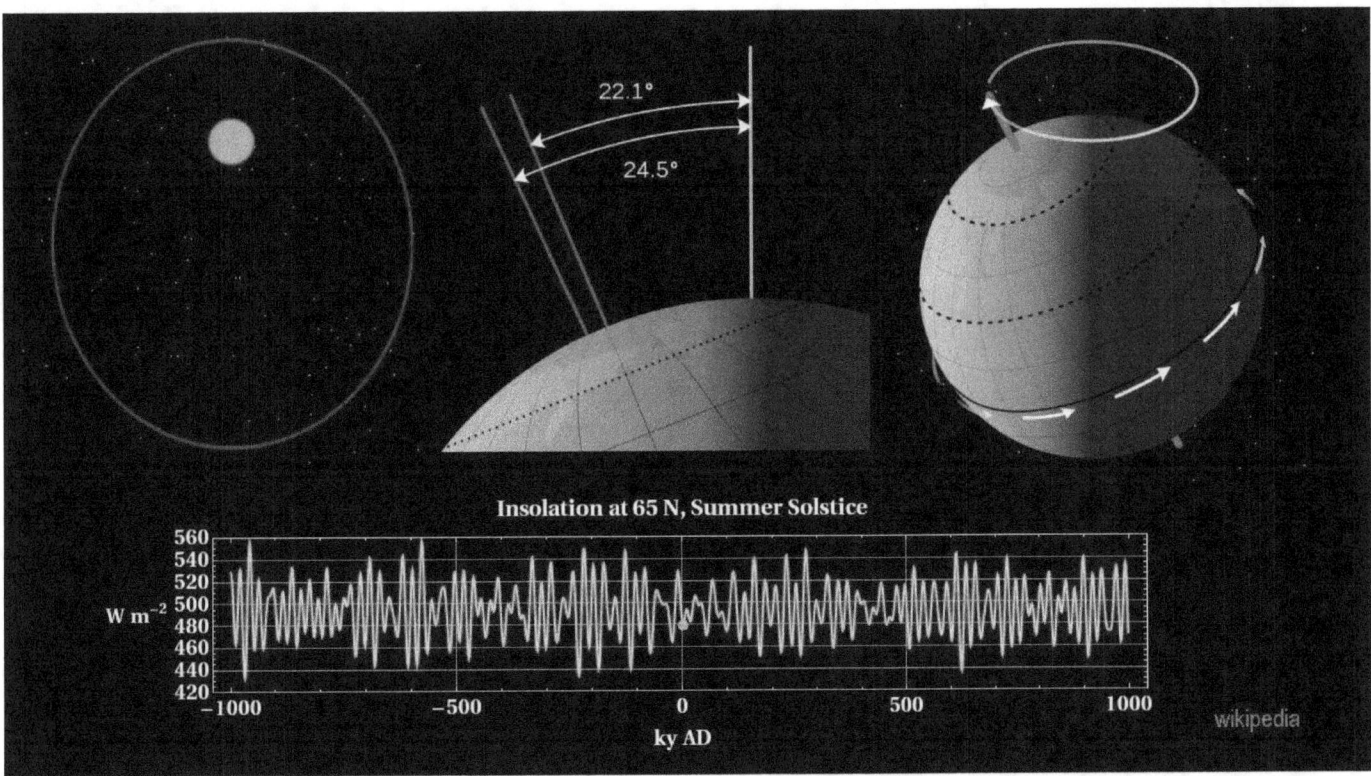

Our civilization is choked with a wide range of false theories that are all pure illusions, and often dangerous illusions for which no supporting evidence actually exists, such as the 'mechanistic Ice Age' theory, where gravity rules everything.

# No truth in any of these theories

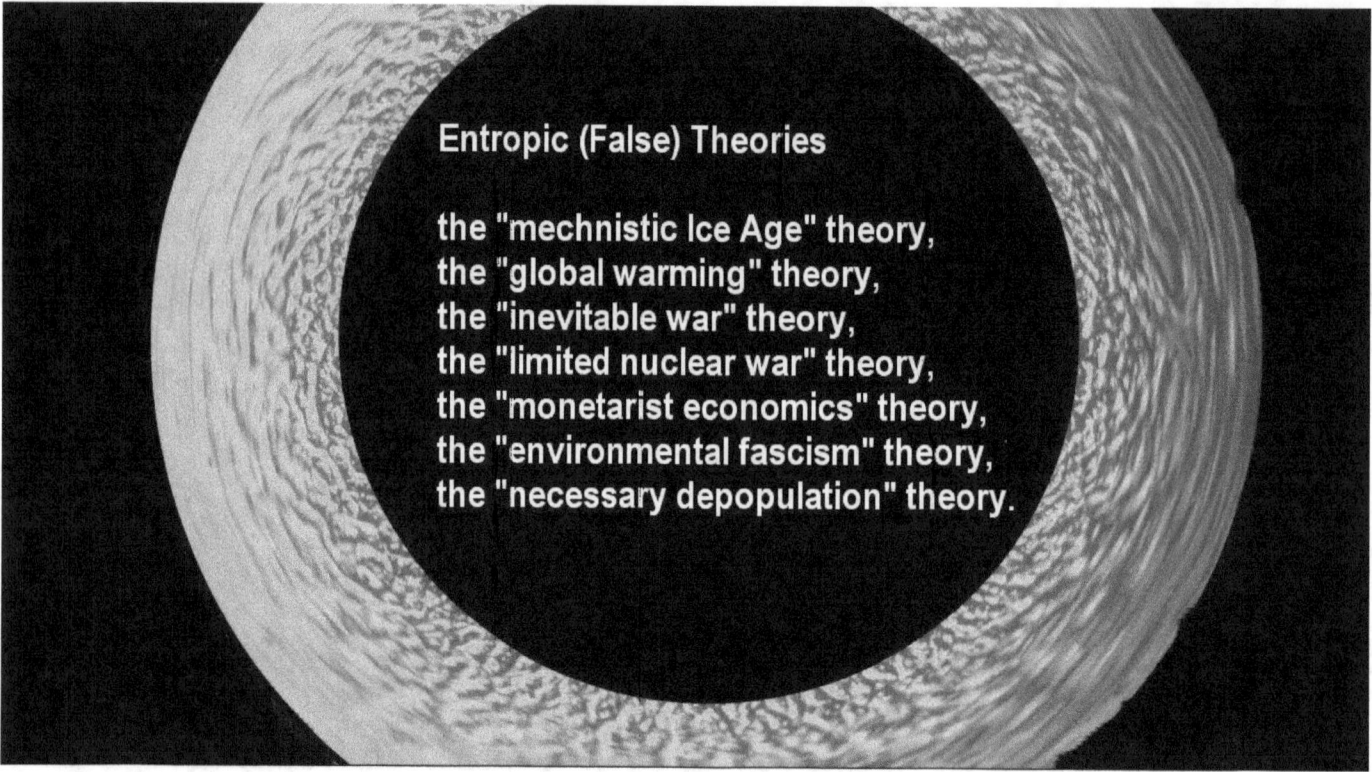

Add the 'global warming' theory, the 'inevitable war' theory, the 'limited nuclear war' theory, the 'monetarist economics' theory, the 'environmental fascism' theory, and not least, the 'necessary depopulation' theory, and so on. There is no truth in any of these theories. These insane types of theories are all built up as false concepts at best, and are typically destructive lies by intention.

# War is not a natural element of humanity

War, for example, is not inevitable. War is not a natural element of humanity that is rooted in the human heart and soul. The opposite is true. War has never benefitted humanity in any way. It is destructive, and therefore entropic. It collapses civilization. It leaves behind an empty landscape, a giant wound that takes a long time to heal.

# Monetarist thievery has never created a nobler and stronger society

Monetarist thievery has the same characteristic. It has never created a nobler and stronger society, but has always collapsed civilization from within.

## limited nuclear war' is a false theory

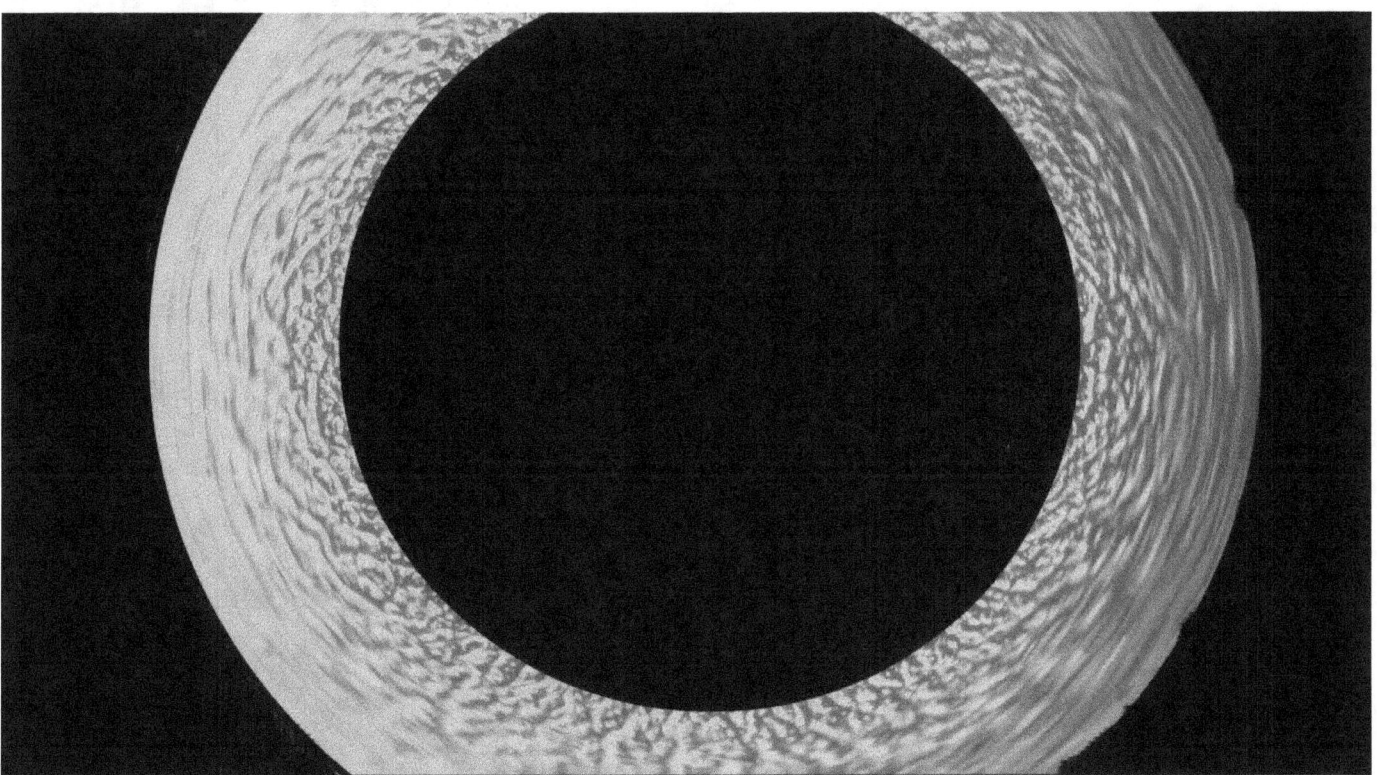

It destroys its own economic center. Neither is the much theorized 'limited nuclear war' possible. It is a false theory. All studies have shown that nuclear war becomes rapidly global, once it starts, with the near-certain extinction of humanity in the wake of it.

# Extinction is the ultimate of entropy

Extinction is the ultimate of entropy. But how do we prevent the ultimate?

# From entropy to anti-entropy

**How do we step away from entropy to anti-entropy?**

With nearly the entire world clinging to entropy as its fate, any practical stepping up to higher ground seems almost impossible. So, how do we do it?

How do we step away from entropy to anti-entropy?

With nearly the entire world clinging to entropy as its fate, any practical stepping up to higher ground seems almost impossible.

So, how do we do it?

# The Pioneering Vision in The Lord of the Rings

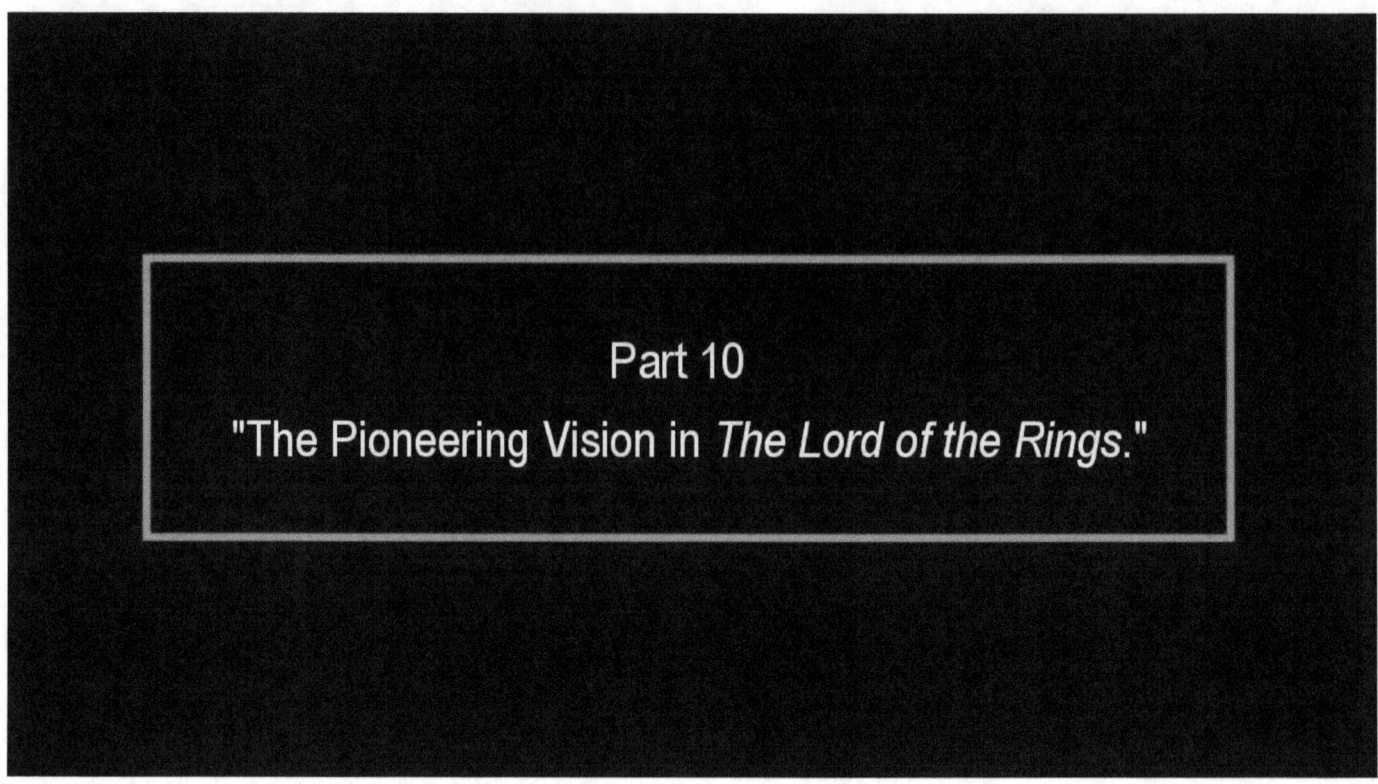

"The Pioneering Vision in The Lord of the Rings."

## Tolkien takes us on a fictional journey

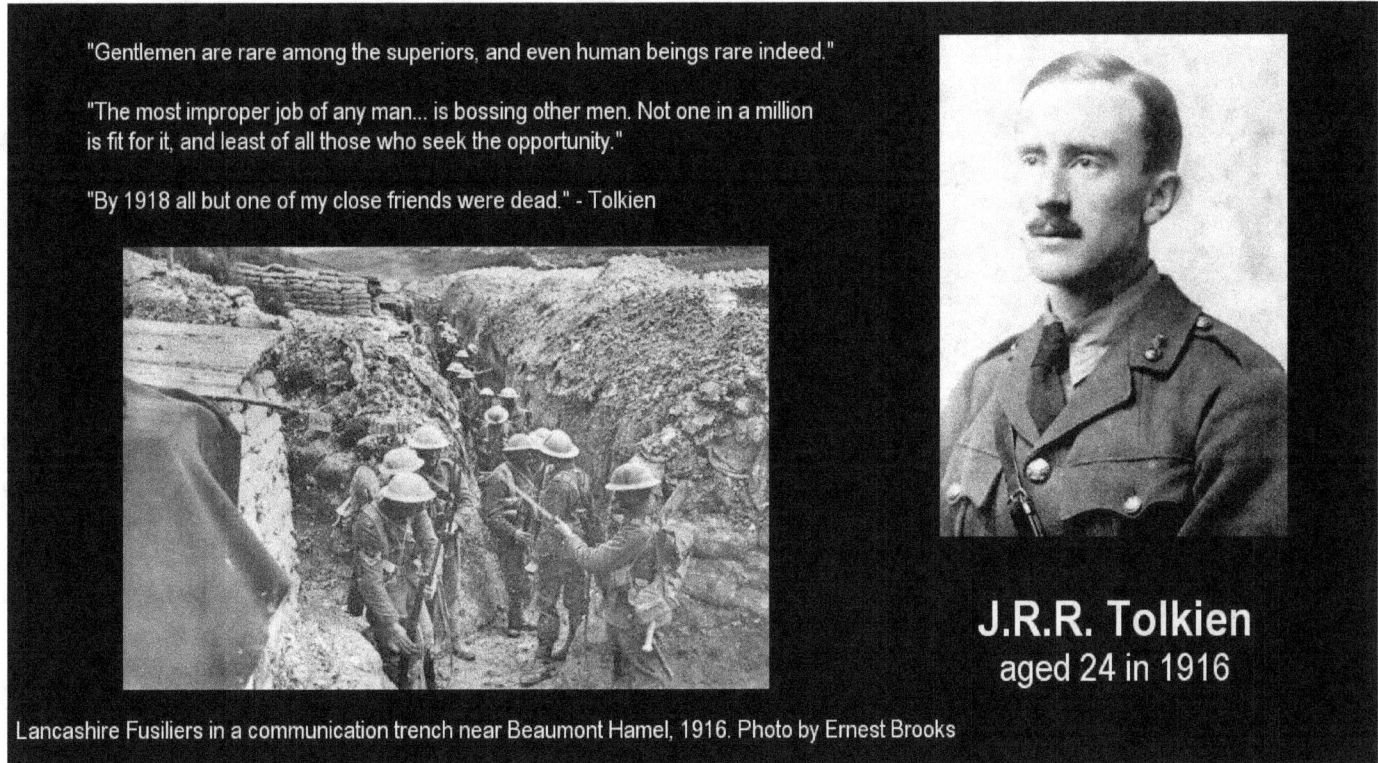

The famous author, J. R. R. Tolkien, explores this very question in his epic saga, "The Lord of the Rings." He explores the path of defending a way of life that already reflects the principle of anti-entropy, but which is threatened to become extinct by the corrupting force of entropy that every system of empire is founded on and serves, which throughout history has stood as the most destructive force in the world, and still does so. Tolkien takes us on a fictional journey in an imagined world to explore the dynamics.

# For the Lord of the Rings

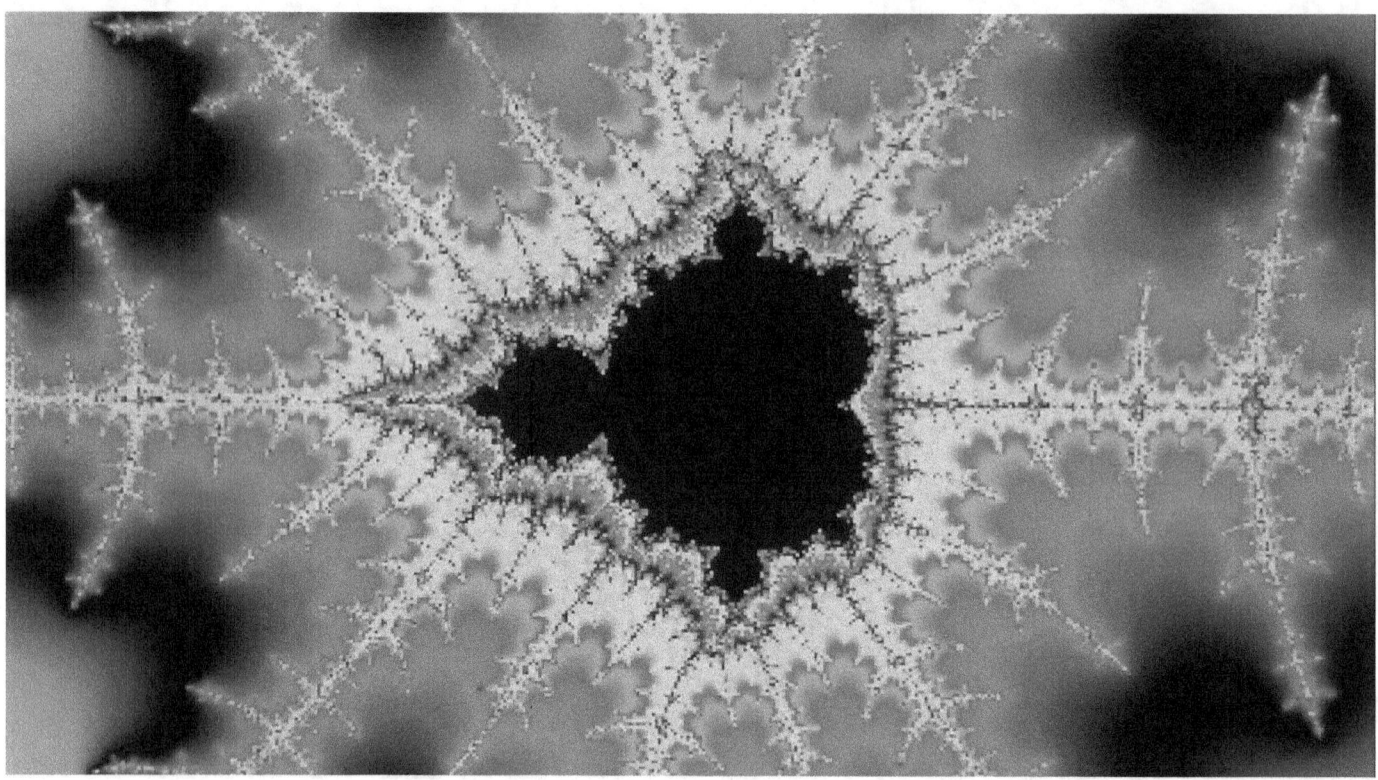

For the Lord of the Rings, Tolkien creates am identifying symbol, the symbol of a ring of fire with an empty center. In the saga, the ring of fire that surrounds nothing signifies the nature of the supreme master of entropy, the master of the high tower of fascism, named, Sauron. The world of men is threatened with extinction by the corrupting force of the fire of the ring of entropy, a type of ring of war that consumes civilization.

Tolkien speaks in metaphor about empire and its effects. In his saga, the ring of fire with an empty center symbolizes the greatest form of evil that has ever established itself, that threatens to end the age of man. The threat defines the battle in the saga, a battle for a society's survival against incredible odds.

Tolkien wasn't far off the mark from the present, in his tale. The modern system of empire seeks the mass-depopulation of the planet from the now 7 billion people living on Earth, to less than 1 billion. The battle in the Ring saga is essentially the same as our battle. But who is fighting the battle in the real world? Who is fighting the forces of entropy?

## Lesser rings, rings of gold

by Jeff Belmonte from Cuiabá, Brazil - Flickr. Licensed under CC BY 2.0 via Wikimedia Commons -

Tolkien also speaks about some lesser rings, rings of gold that have corrupted many, with one special ring among them, a "master ring," which the master of the empire of evil had once on his finger, but which became lost for a long time. As a symbol, Tolkien shrank the great ring of fire with an empty center into an object that a person can wear, a golden ring with an empty center that drains a person's humanity away, who thereby becomes a slave to it. The corrupting power of the "master's ring" is so intense in the tale that any wearer of it becomes invisible to everyone around in the normal world as if the wearer did no longer exist. That's what happens in the real world to those who surround themselves with the corrupting influence of entropy. In the saga, the master ring is also called "my precious."

This is the type of effect in the real world that the Big Bang theory has on all who 'wrap' the lie of entropy around their finger. These are they whose contribution to science has essentially vanished as if they had exited the universe, or had never lived. To the people stuck in this trap, the Big Bang Cosmology is everything from A to Z in the universe, who doom the world thereby. In the saga, for as long as the 'master' ring exists, the world of man is likewise doomed. But Tolkien makes it clear that the 'master' ring, though it must be destroyed, cannot be destroyed with any craft one might employ. A corrupting false theory cannot be destroyed with an axe, but must be returned to the chasms of evil where it was forged. It needs to be traced back to the chasms of empire, the chasm of the fire of lies.

## Casting the theory of entropy into oblivion

"Lava Lake Nyiragongo 2" by User:Cai Tjeenk Willink (Caitjeenk) - Own work. Licensed under CC BY-SA 3.0 via Wikimedia Commons -

Tolkien assigned the critical task of casting the theory of entropy into oblivion, to a few pioneers of the society who already live by the principle of anti-entropy in all they do. They live it, by living a rich life with abundant resources that they have created for themselves, and continue to do so. Tolkien placed the tallest task on these few.

## Tolkien leaves no room for a compromise

"Stonehenge Summer Solstice eve 02". Licensed under CC BY-SA 2.0 via Wikimedia Commons -

"One of you must win this necessary fight against entropy," a wizard proclaims to a council. Tolkien selects one of the few who already live the principle of anti-entropy, and selects the smallest of them, as if he was saying to him, "only you can do this. You know in your heart what is true. You have lived the principle. The others who are standing on a lesser platform don't have a chance. One cannot win against an evil by fighting on the low corrupting level where the evil has been forged. This doesn't work. But you stand on a higher level. That's why you can win, who are living on that higher level where the supreme principle is experienced as real. This means that You must do this. You have qualified yourself to succeed." Tolkien assigns a wide range of lesser figures in the tale, to help the project to succeed." And they all, cooperating and contributing all that they have, do succeed against the most incredible odds in a fight that spans three volumes in book-form.

Tolkien leaves no room in the story for a compromise with the Empire of Entropy that is poised to destroy the world of man. It is tempting of course, to reach for a compromise, such as to use the supposed 'power of entropy' against its empire who hails it, to defeat it. Such attempts are indeed made in the story, all the way through, and they all fail and result in tragedy.

Tolkien puts the world on notice, not to fail. Still, it did fail in the real world.

## The "Glass Steagall" banking legislation

1933 - the Glass-Steagall law to protect that value of the U.S. currency - now repealed

A great attempt had been made to save the USA from the ravishing of the Empire of Entropy. It is known as the "Glass Steagall" banking legislation of 1933. It was designed to protect the nation. However, it was also built on a compromise. Tolkien's tale was written between 1937 and 1949, with the bulk of it during World War II. More than 150 million copies have been printed of the tale. Nevertheless, its fundamental premise has remained unrecognized. The compromise in the Glass Steagall legislation in the USA, has never been addressed, much less overturned, whereby the law itself was eventually defeated, and the protection it had afforded, was lost, with the result of great tragedies for the nation and the world. The tragedies still continue.

# Among the foes in the Lord of the Rings

Foreground, left to right: Führer Adolf Hitler; Hermann Göring; Minister of Propaganda Joseph Goebbels; Rudolf Hess

Among the foes in the Lord of the Rings, Tolkien places a group of black-glad kings who were once men, but who became corrupted into becoming creatures without a face and without a soul, who wield destructive power, as they all serve Sauron, the foulest evil of them all whose symbol is the ring of fire with no center.

## We have many evil potentates in high places

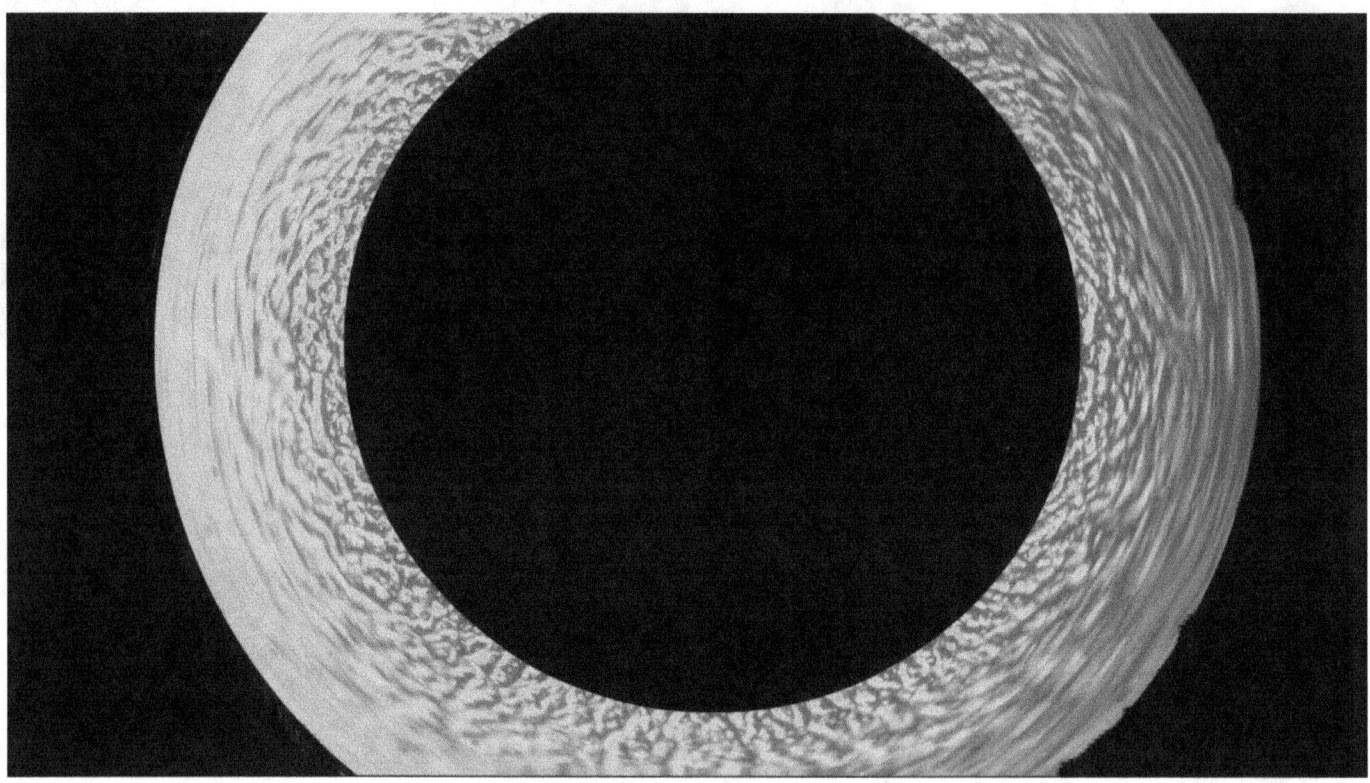

We have many such evil potentates in high places in the world to whom the symbol of an empty center surrounded by a ring of fire, would apply.

## The supernova war that nothing survives

Those are the criers for war, even for the big bang of nuclear war, the supernova war that nothing survives.

## Masters of false theories

They are masters of false theories. Nuclear war is a false theory. Every form of nuclear war is a false theory, even limited nuclear war that is presently prepared for to eradicate Russia, China, and India.

# Russia and China placed at the cross-hairs

We see Russia and China placed at the cross-hairs for a surprise limited nuclear war, by scores of empty people who have lost their qualification to be called human beings.

# Where people lie to themselves

USA - Ohio-class strategic missile submarine - (14 in service)

In their land, where people lie to themselves, the world-engulfing conflict of limited nuclear war is miraculously deemed survivable in some fashion, for which the stage has been set up, and for what?

# When society looses its renaissance of the truth

by Rembrandt
(1606–1669)

When society looses its renaissance of the truth, insanity rules the landscape, whereby the most precious we have on the Earth, which is humanity itself, is doomed to suffer extinction.

# The American Paradox

"The American Paradox"

## America stands as a paradox

America stands as a paradox that has not yet been resolved. It was founded as a republic. A republic is inherently anti-entropic in nature. When forms of government by a society of itself, are focused on increasing the wealth of its living, and thereby its freedom from limitation by the creative and productive power of its humanity, then the resulting system of economics is anti-entropic in nature. It is so, because the wealth of its living, from food to housing to transportation, is increased by the society's scientific and technological progress, and by its commitment to empower this real wealth-building process by all means possible.

In the resulting environment society functions by supporting one-another on the entire front for its common welfare. The dynamic process includes quality education, health care, increased beauty, expanding love, and an elevating culture. On the recognition of this universal-freedom platform many of the former colonies in North America eagerly joined hands to became a nation.

## Why has the nation fallen

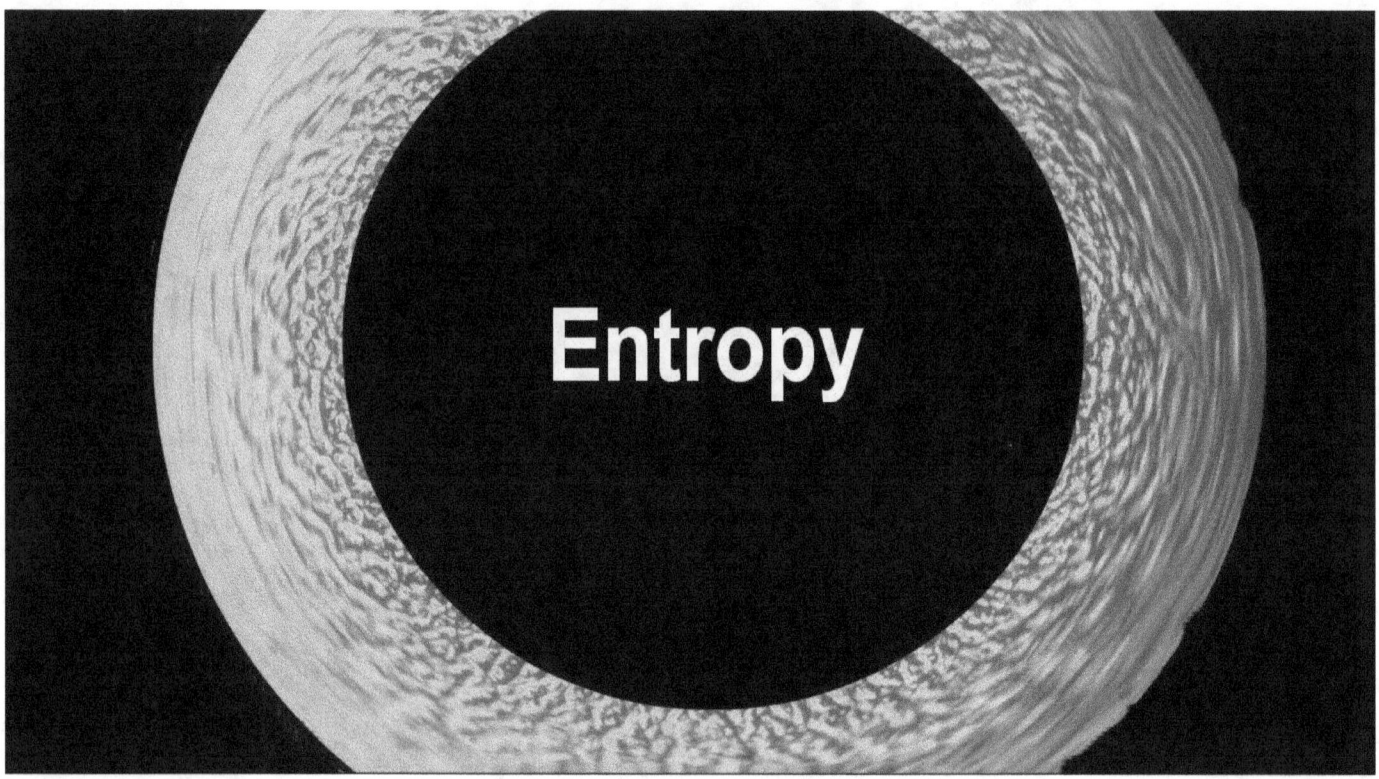

But why has the nation fallen from the high platform it has been built on? Why has it fallen back to the level of the kingdoms of entropy where human value decays?

## The universe doesn't fall into entropy

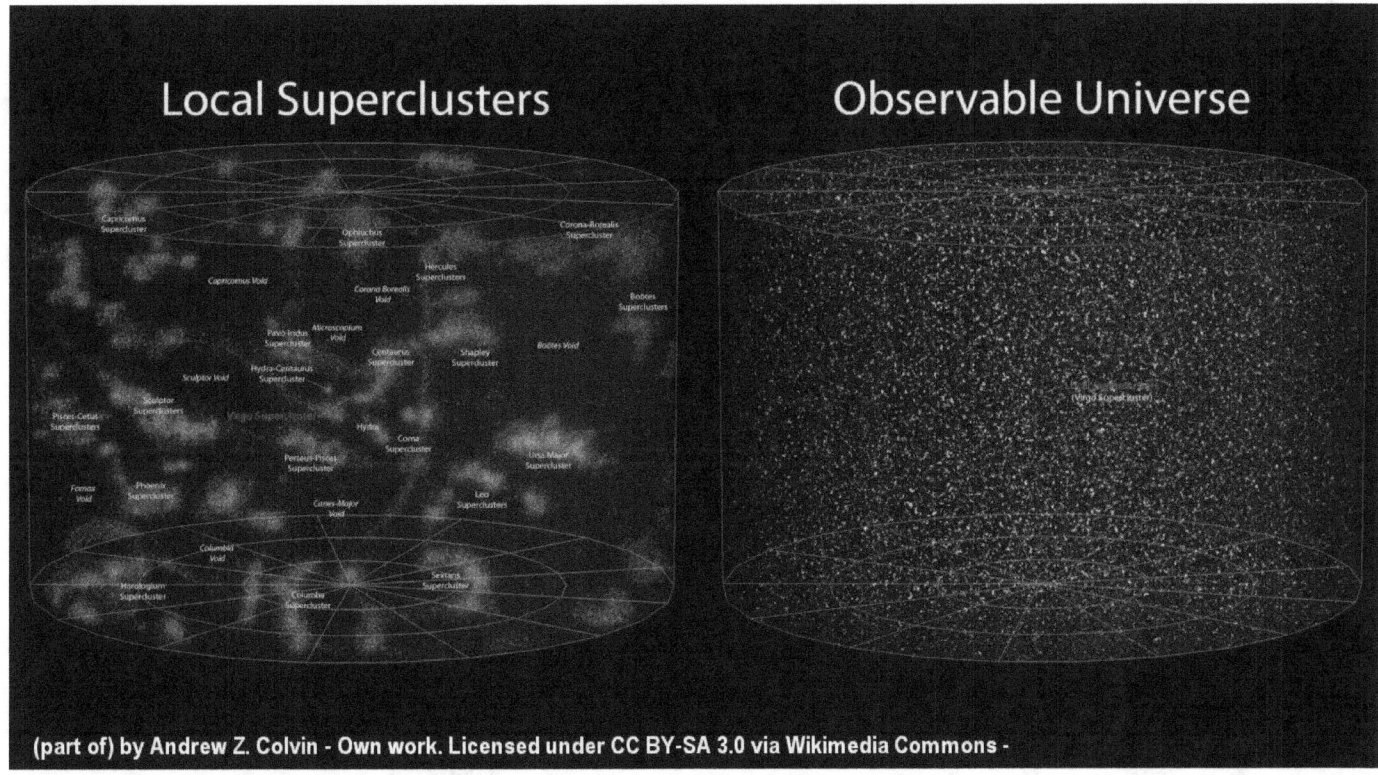

(part of) by Andrew Z. Colvin - Own work. Licensed under CC BY-SA 3.0 via Wikimedia Commons -

America should not have fallen. The universe doesn't fall into entropy. The universe operates on the anti-entropic platform where nothing stands in isolation and is self-consuming, where instead everything is actively powered by a process in which every single star of every galaxy plays a contributing role. Without this contributing, active participation, in the universally creative process that the universe is, with its myriad expressions, not a single star would exist in cosmic space, which would thereby be an empty void.

## America the paradox of a fallen star

America stands as the paradox of a fallen star, because a portion of society believes that it can exist and prosper on the opposite platform than that of the universe. The opposite platform, in this case, is the platform of entropy ,reflected as the lie in economics that invites the tragedy of stealing from one another, such as by slavery, deceit, greed, and other forms of larceny, even rape, such as by waging war for stealing.

America stands as a paradox, because it aims to exist on two opposite platforms: the platform of Entropy where stealing is King, and the Anti-Entropic platform were society's creative self-development is pursued and is powered by its humanity and its human resources. The conflict between the two systems has shaped American history.

American history is a history of great achievements compromised with great tragedies, a history of America being a light into the world, and of it also being the greatest force of fascism and war that stands poised to extinct humanity on the pathway of universal stealing.

## The paradox that America became

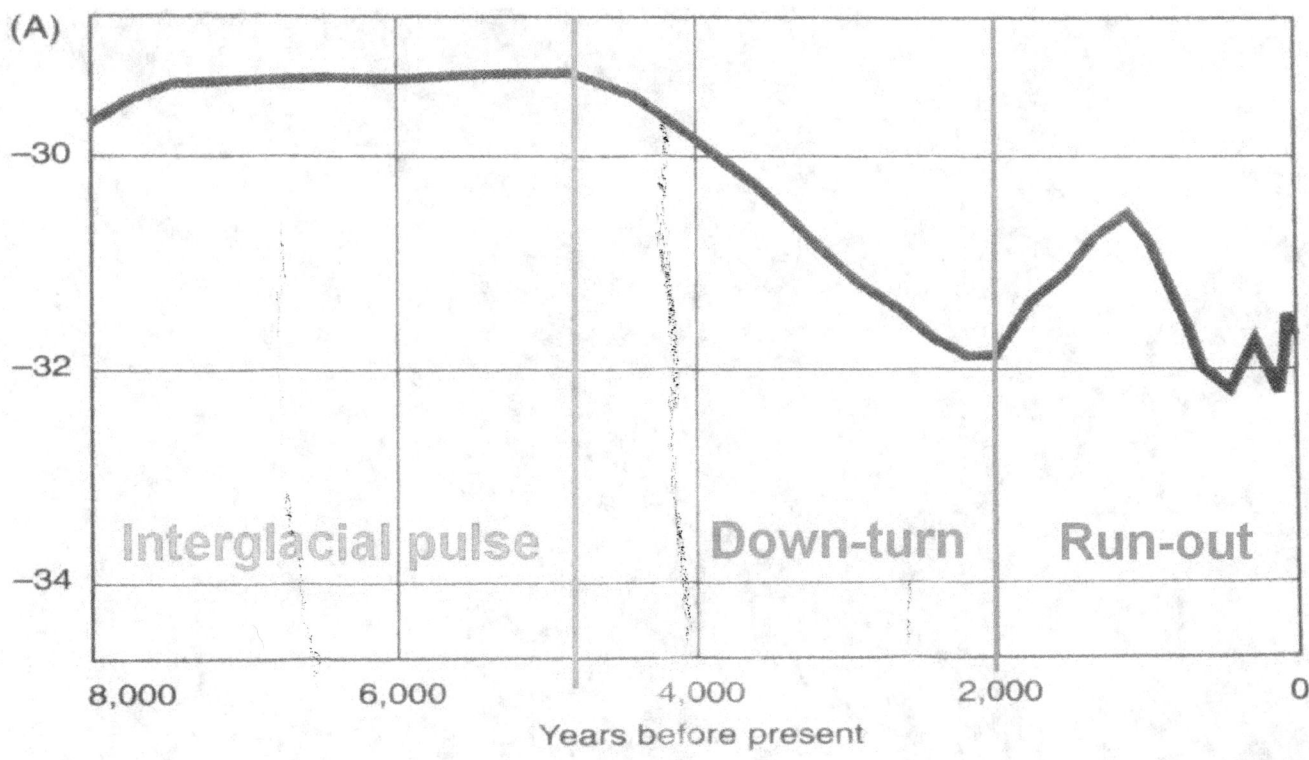

The paradox that America became, appears to have started millennia before its time when the climate of the Earth began to cool from the warm climate of the interglacial optimum. In the shadow of this climatic down-turn, the Earth's primitive biological system became less productive. Food became less abundant. The idea might have emerged that wealth in living can be increased by stealing, which leads to wars for stealing. In the less-productive landscape roving bands may have ravish the lands. Some of the bands became empires - empires that are trapped by the notion of entropy that invites stealing. This is the trap that society fell into, by its small-minded thinking. This deadly trap in thinking still chokes much of humanity.

## One of the great masters of the trap

One of the great masters of the trap was Adam Smith, who is hailed the father of modern economics that should be termed, mistaken economics. His take was that economics is the result of individuals serving their self-interest, led by the invisible hand of greed, acting in disregard of the public good.

Smith trapped himself in this prison of entropy where the fierce heartbeat of relentless competition diminishes all human value to zero and elevates money value to be the all-precious. Thus greed becomes a deity that demands the sacrifice of everything else. Adam Smith calls this economics, even while he 'defies' the principle of economics.

# Adam Smith was mistaken

Of course, Adam Smith was mistaken. The hand of self-serving greed that acts in disregard of the universal good, is not invisible in the real world, it never has been, but has always been ravishing. Adam Smith stands self-condemned thereby, by erecting a throne that stands opposed to the anti-entropic platform of human development where creativity, science, and productivity are rooted. His throne opposes the only source of true economic value that exists. He opposes the system of humanity where stealing is NOT required, but is regarded as a crime.

The process of stealing as an economic concept, which the empire of western economics has become, is inherent only in the imagined entropic model that emerged out of the ages of small-minded thinking. That's what Adam Smith had idealized, a model of relentless competition for the gold in which human value falls by the wayside.

## When money is an object for stealing

> **The Smith model is immensely destructive**
>
> When 'money' is for stealing,
> wars are waged for increased stealing.
>
> There has never been a war instigated in all history,
> that wasn't for the purpose of stealing. Wars are for stealing.

The Smith model is immensely destructive.

When money is an object for stealing, wars are waged for increased stealing. It is useful to note here that there has never been a war instigated in all history, that hadn't been for the purpose of stealing, fundamentally. Wars are for stealing, exclusively.

## In the extreme case, as we have it today

In the extreme case, as we have it today in the disintegrating world of imperial monetarism, the resulting extreme desperation for continued stealing opens the gates to the unthinkable, to nuclear war, the ultimate 'competition' that no one survives.

## Nuclear war, in any form

> nuclear war,
> in any form,
> **is unsurvivable**

Nuclear war, in any form, is unsurvivable.

# A dead peace without a human voice

Nuclear war will lead to a dead peace without a human voice, or any voice at all.

## Far distant from Adam Smith's economics

Mumbai Bridge

The potential future of humanity lies far distant from Adam Smith's economics that still drives the western entropic platform for stealing and destroying, which the developing world is fast moving away from in the emerging cooperative commitment by Russia, India, and China, towards a world of building for the universal welfare of society.

# The future of humanity

India Mumbai Bridge (Wikipedia) — Amit Kulkarni

The future of humanity lies in rapidly increasing the wealth in human living. It lies in moving away from destroying humanity, such as by starvation, war, and depopulation, to developing the creative freedom and unlimited productivity that humanity is capable of.

# Real economics aims for the goodness in living

"Taj Mahal by Amal Mongia" by amaldla from san francisco CC BY-SA 2.0 via Wikimedia Commons

Real economics aims for the goodness in living, without stealing, without greed, without war, without destroying one-another in the grasping for gold, but with evermore building, uplifting culture, and beauty. The future of humanity lies with the principle of anti-entropy, the principle of the universe, reflected in the unfolding creative power of humanity.

# After India became a free nation in 1947

The Lotus Temple, located in New Delhi, India, a Bahá'í House of Worship completed in 1986. 70 million visitors by 2001

"LotusDelhi" by Vandelizer - http://www.flickr.com/photos/jeremy_vandel/99170173/sizes/o/ Licensed under CC BY 2.0 via Wikimedia Commons -

After India became a free nation in 1947, that ended its long history of being looted in numerous colonial processes by numerous masters, in which many tens of millions of people have perished, By claiming is freedom from the thievery India gave itself a chance to live again, which hadn't been possible during the dark centuries of it being ravished by foreign masters who had subjugated most of the world, as they did India.

# Segment 4 - Colonial Age & Glass Steagall Compromise

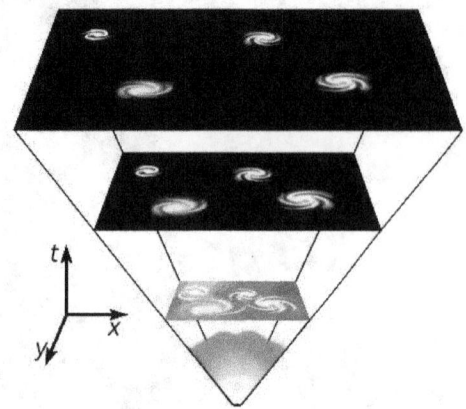

# Big Bang Blow-Out

## Segment 4 - Colonial Age & Glass Steagall Compromise

Both colonialism and America's attempt to free itself from it with the Glass Steagall law are world-historic failures wrought by the compromise that upholds entropy in civilization. Colonialism failed thereby, as did Glass Steagall in America. Both attempts to compromise were immensely tragic for the nations of the world, and still are, as in the case of the 'Euro Colonial System.'

Both the colonial system and the Glass Steagall legislation, were attempts to shield civilization from the destructive effects of the built-in entropy of the imperial, oligarchic system. The goal should have been to eradicate the entropic systems. Society' submitting itself to them, in the form of a compromise, is comparable to it inviting rape upon itself. The result is tragic in every case. The Big Bang theory has the same effect, as a platform with an empty center.

# The colonial rule for stealing the wealth of nations

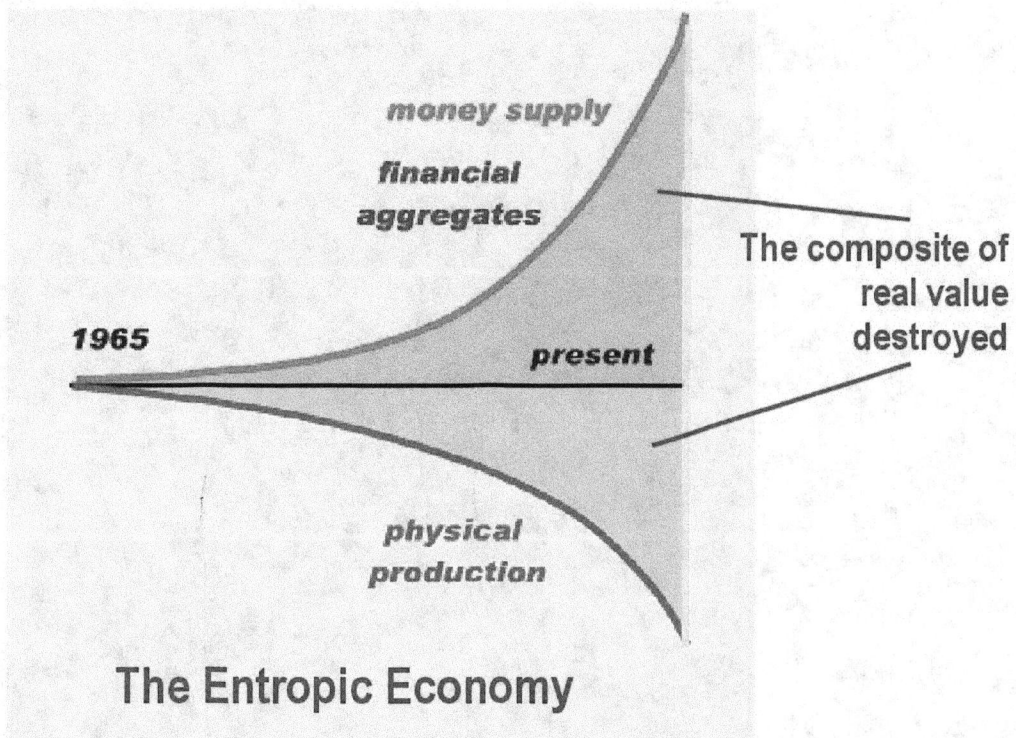

The colonial rule for stealing the wealth of nations, does actually precede 'Adam Smith', who may have tried to hide it. It appears to have been recognized quite early in the history of civilization that the process of stealing, in whatever form it may have been carried out, collapses the thereby ravished economy, so that the stolen wealth itself becomes worthless thereby for the simple fact that monetary values stand as a claim against a productive economy, which under the regime of looting becomes a destroyed economy.

# The Colonial Age,

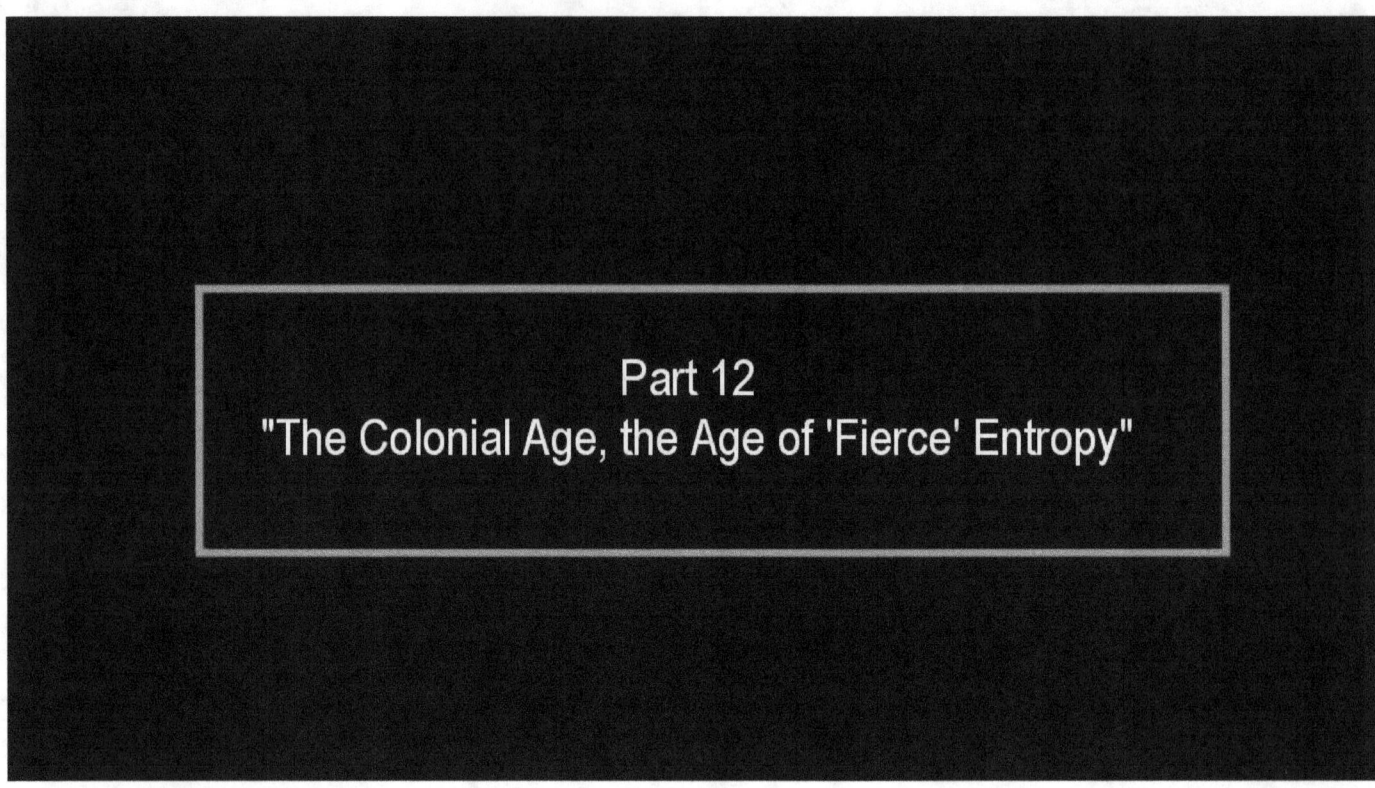

"The Colonial Age, the Age of 'Fierce' Entropy"

# To circumvent the inherent entropic collapse

It may have been in the attempt to circumvent the inherent entropic collapse that is an unavoidable feature of the thievery system, that the masters of the greedy began to spread their thievery across the world to other nations, to ravish them instead of their own local society that the masters found themselves obliged to maintain as a necessary infrastructure for its wars.

# The colonial age began

With their success in spreading their looting 'enterprises,' the colonial age began, which of course was enforced with the infamous gunboat 'diplomacy,' and if there was resistance, with war to enforce the raping.

# When the resistance was internal

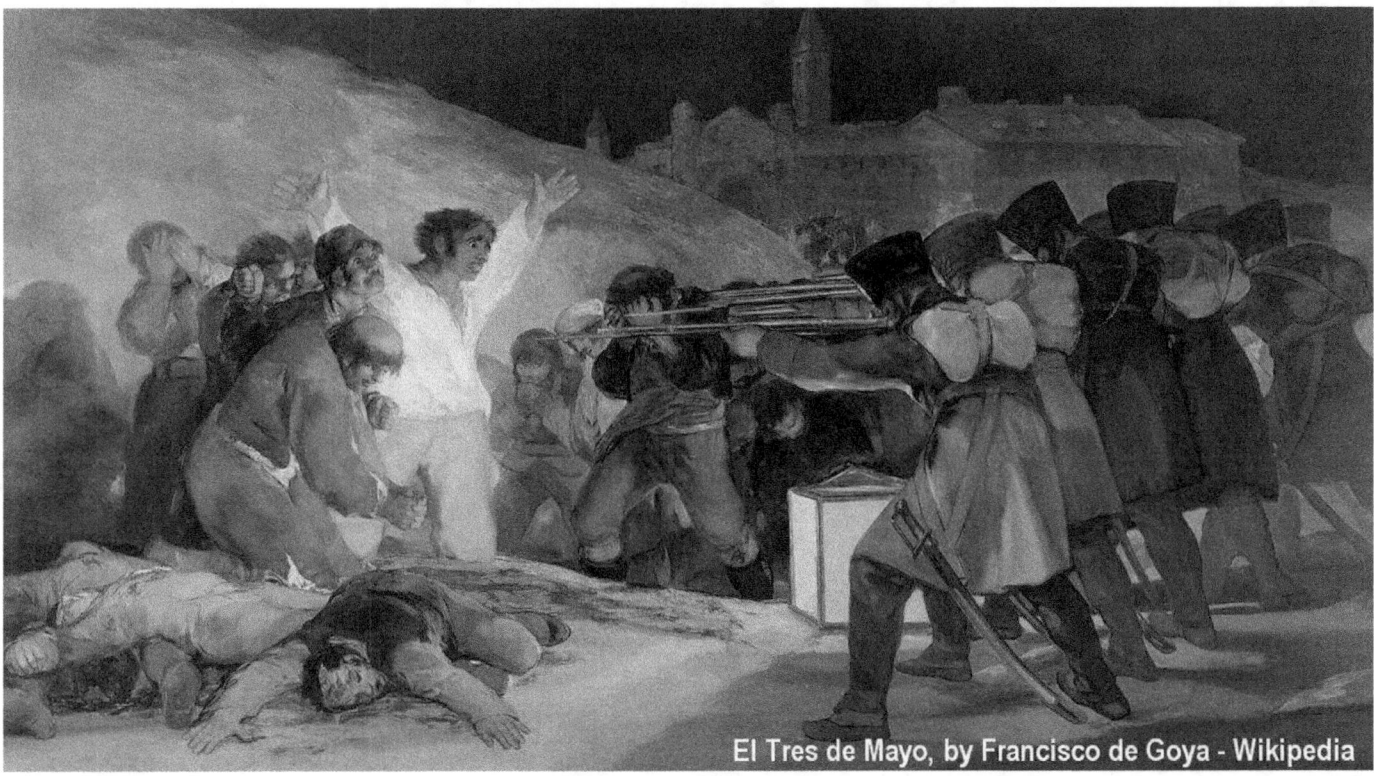

El Tres de Mayo, by Francisco de Goya - Wikipedia

And when the resistance was internal, genocide was unleashed to destroy the hart of of whatever society had dared to raise it head in defiance.

## Almost the entire world became subjugated

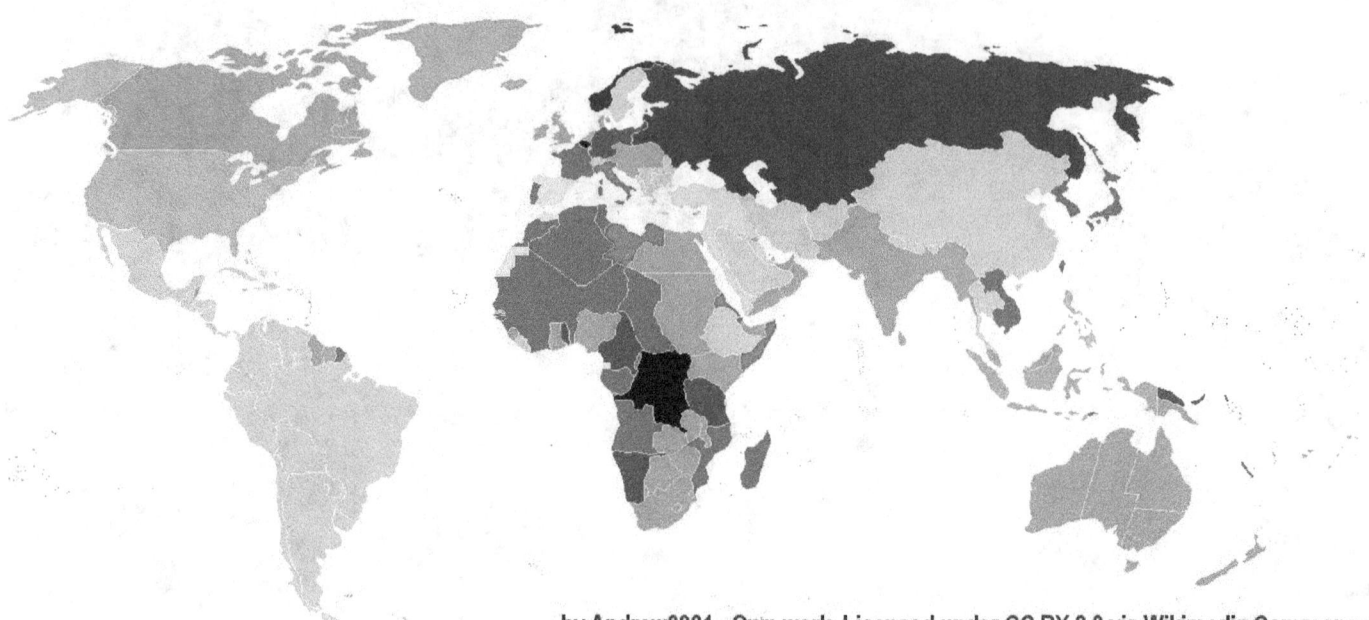

Colonial Empires of the World 1914

by Andrew0921 - Own work. Licensed under CC BY 3.0 via Wikimedia Commons -

During the colonial age, almost the entire world became subjugated in various ways to tiny groups of oligarchic money bags and royal houses in the 'empire lands,' the lands of the private masters who have looted the world far and wide for their pleasure, by whom humanity was treated as a type of animal to be 'harvested' in the extreme, and discarded at will.

Segment 4 - Colonial Age & Glass Steagall Compromise

# Lists of the subjugated nations

| British colonies | | French colonies | |
|---|---|---|---|
| Aden | Gilbert and Ellice Islands | Algeria | Upper Volta |
| Anglo-Egyptian Sudan | Gibraltar | Clipperton Island | Guadeloupe |
| Ascension Island | Gold Coast | Comoros Islands (including Mayotte) | Saint Barthélemy |
| Australia | India | French Guiana | Saint Martin |
| Australian Antarctic Territory | (including Pakistan and Bangladesh) | French Equatorial Africa | La Réunion |
| Christmas Island | Ireland | Chad | Madagascar |
| Cocos Islands | Jamaica | Oubangui-Chari | Martinique |
| Norfolk Island | Kenya | French Congo | French sMorocco |
| Bahamas | Malta | Gabon | New Caledonia |
| Basutoland | Newfoundland | French India | Saint-Pierre-et-Miquelon |
| Bechuanaland | New Zealand | French Indochina | Shanghai French Concession |
| British Antarctic Territory | Cook Islands | Annam | Tunisia |
| British East Africa | Niue | Cambodia | Vanuatu |
| British Guiana | Ross Dependency | Cochinchina | Wallis-et-Futuna |
| British Honduras | Tokelau | Laos | Not shown here: |
| British Hong Kong | Nigeria | Tonkin | Russian colonies |
| British Malaya | North Borneo | French Polynesia | German colonies |
| British Somaliland | Northern Rhodesia | French Somaliland | Italian colonies |
| Brunei | Oman | French Southern and Antarctic Lands | Dutch colonies |
| Burma | Papua | French West Africa | Portuguese colonies |
| Canada | Sarawak | Benin | Austro-Hungarian colonies |
| Ceylon | Sierra Leone | Côte d'Ivoire | Belgian colonies |
| Cyprus | Southern Rhodesia | Dahomey | U.S. colonial possessions |
| (including Akrotiri and Dhekelia) | St. Helena | Guinea | Chinese colonies |
| Egypt | Swaziland | French Sudan | Ottoman colonies |
| Falkland Islands | Trinidad and Tobago | Mauritania | Japanese colonies |
| Fiji Islands | Uganda | Niger | |
| Gambia | South Africa | Senegal | |

The lists of the subjugated nations, countries, and areas, were long. They contained many names of countries and people and areas that became 'property.'

On this page, only the names of the two biggest groups of colonial properties are listed. These are the names of the British colonies, and of the French colonies. The remaining eleven colonial owners are listed without their properties displayed.

Countless wars were fought throughout the centuries to subjugate the colonies, and to keep them in line.

## The American republic was born

The American republic was born when a group of colonies took a stand against being subjugated by the global thievery system. The American patriots claimed the inherent freedom of the human being for themselves. They claimed their freedom on the recognition of the inherent anti-entropy in the human society that came to light through its scientific and technological development on the distant shores of America, far from the choking effect of Empire. The patriots discovered that a society's inner development invariably increases its creative and productive power. The discovered anti-entropy of their humanity gave the patriots the dignity with which they stood tall, and said to empire: No More, Never Again.

# America did attain its freedom, and fought to retain it

Washington Crossing the Delaware, in 1777 - "These are the times that try men's souls," (Thomas Paine)

On this basis America did attain its freedom, and fought determined to retain it against all the forces of empire that promptly waged a war to reverse the unfolding history of humanity.

America stood its ground for 130 years.

## America lost itself again to the devil within

America lost itself again to the devil within, to the theory of entropy, to the platform for stealing that invited the Federal Reserve Act, the platform for war. It never freed itself from this defeat, but was drawn into world wars by the process.

## Colonial wars soon became world wars

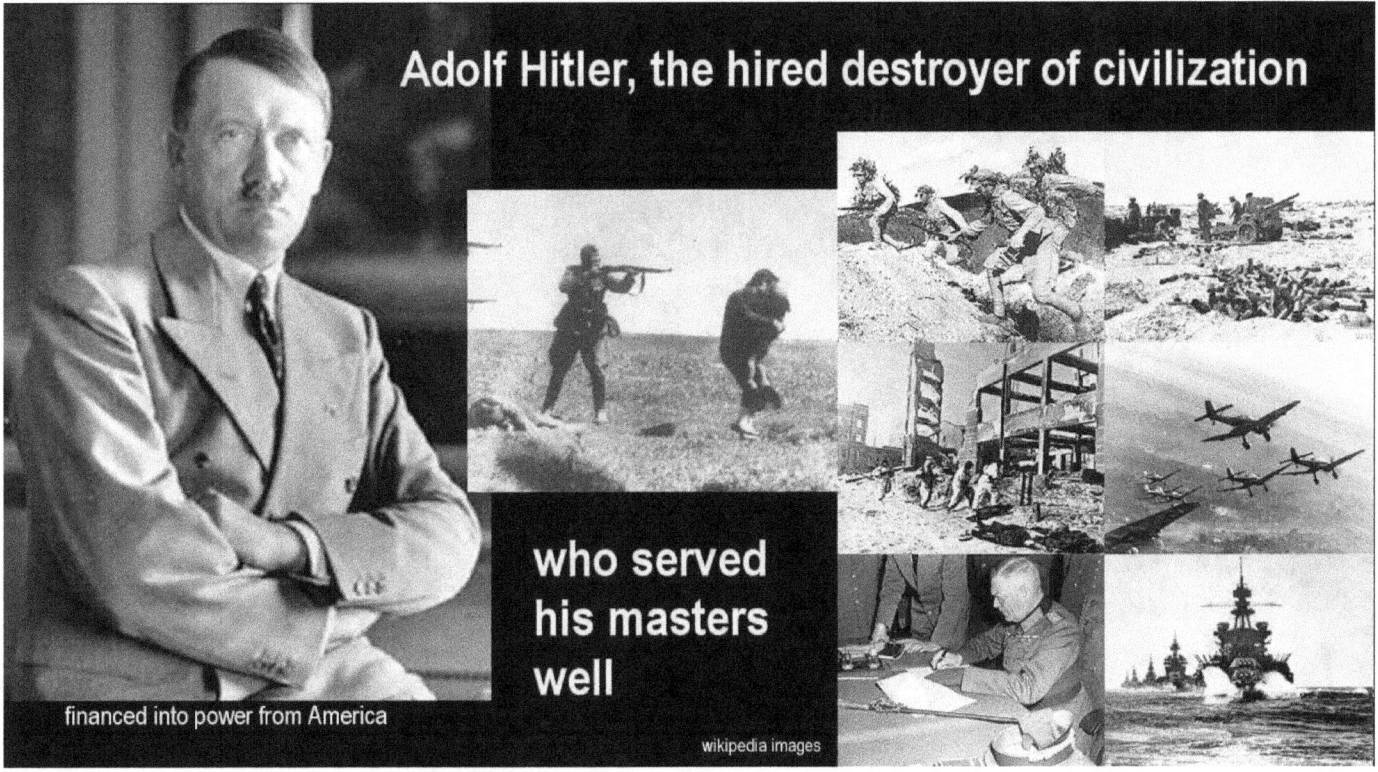

The historic colonial wars soon became world wars, poised to become nuclear war. The buzz words, "First Strike," "Limited," "Pre-emptive," are now applied to nuclear war against the whole world, just as Hitler had applied these terms to his grand schemes of madness in obedience to his masters who had set him up as their puppy dog.

## Stealing with the force of 'invincible' arms

While the face of war has been radically modernized in recent years, the platform for war remains the same nevertheless. The masters who stand behind all the platforms for stealing, cling tenaciously to their mistaken belief in universal Entropy for which their devil in the mind inspires stealing with the force of 'invincible' arms, while the masters themselves never go to war to die in the conflicts they create. Soldiers are used for that.

# Few soldiers in history knew

UH-1D helicopters airlift members of a U.S. infantry regiment, 1966 - James K. F. Dung, SFC, Photographer

Few soldiers in history knew, or even wished to know, that they laid down their life to die in the dust of some foreign lands for the profits of the wealthy who own them on their strings, and who thereby render them to become mass-murderers before God in the wars that demand them to discard their humanity as a worthless impediment.

## War has become a worldwide disease

Russian made HIND Mi-24 helicopter

War has become a worldwide disease that is waged for the purpose of looting, for which enormous resources are wasted in the defence against it.

Sadly, in the heat of the raging battles, the root for the disease, the belief in entropy, becomes largely forgotten. Only the blood remains real that flows into the sand.

## Stealing by all means possible, humanity is doomed

For as long as the historic, entropic game of empire continues, which demands stealing by all means possible, humanity is doomed to its self-destruction by war. Without science raising up the sovereignty of our humanity to higher levels, above the small-minded historic game, our hope is slim.

## As stealing demands evermore wars

For as long as stealing demands evermore wars, humanity denies itself to have a human future.

# The whole of humanity is now doomed

The whole of humanity is now doomed by its belief in entropy that invites stealing. The result is exceedingly tragic. The human world cannot survive the force of half a million Hiroshimas.

## Their greed demands, a scientific excuse

With the formulation of the cosmic Big Bang theory, the masters of the system of empire attempted to give their deadly and destructive system of entropy, a noble face to justify their terror. With the Big Bang theory, the masters attempted, not to heal their inhumanity, but to give the increasingly horrific destruction of civilization, that their greed demands, a scientific excuse.

## Science complied

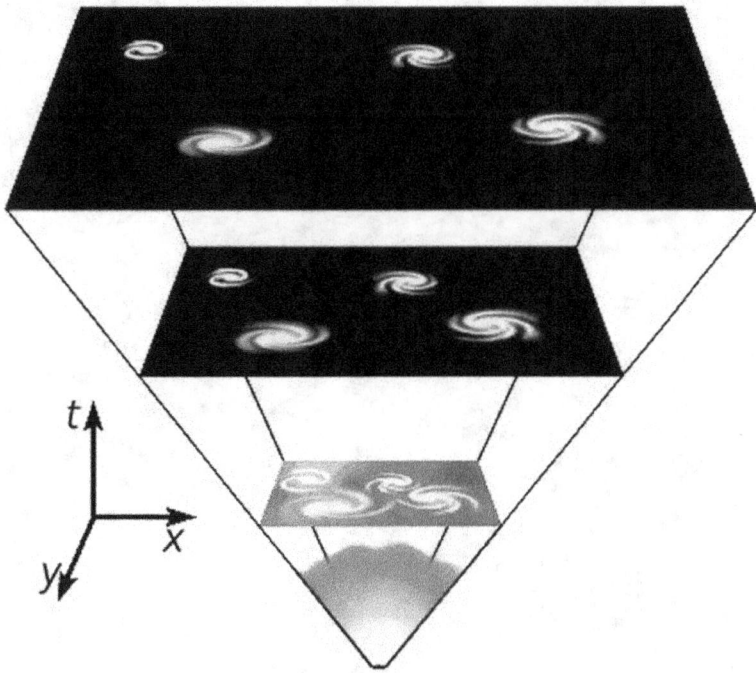

Science complied. It pinned the ugly face of empire, the face of entropy that trails out into nothing, unto the face of the universe, even unto the face of God.

That's an old trick that the empire of Rome had used, and all empires thereafter under the rubric of the "divine right of kings."

While the result, that became the cosmic Big Bang theory, is evidently false, as no real evidence supports the theory of the declining and self-consuming universe, the theory's myth of universal entropy nevertheless still rules society powerfully. It rules it as a philosophy that upholds the dream of entropy, a dream where all energy is consumed into nothing, which is the key feature that the world empire cannot avoid, with it being built on the platform of entropy.

# Thievery inherent in the kingdom of empire

The disease of thievery as a system that is inherent in the kingdom of empire, has unfortunately become so powerful by it's being promoted as a science, so that the budding opposite political structures that are based on the platform of anti-entropy, the platform of human self-development and self-protection, have become largely eliminated from the landscape of civilization in the modern, neo-colonial, imperial world, for as far as its tentacles have been able to reach.

# A new wind is rising in the distant lands

Fortunately the disease is fading. A new wind is rising in the distant lands where the tentacles of empire are loosing their grip. The winds of healing are rich with scientific development, cultural optimism, infrastructure building, industrial production, energy development, cultural development, which are all features of human development. These winds flow from China today, and Russia, and India, who have become the pioneers for a new hope for humanity.

# Where America had stood when it stood tall

These pioneering nations, and those who joined hands with them, stand at the forefront today in the race away from empire, where America had stood when it stood tall, before it became re-colonized again.

# The Glass Steagall Compromise

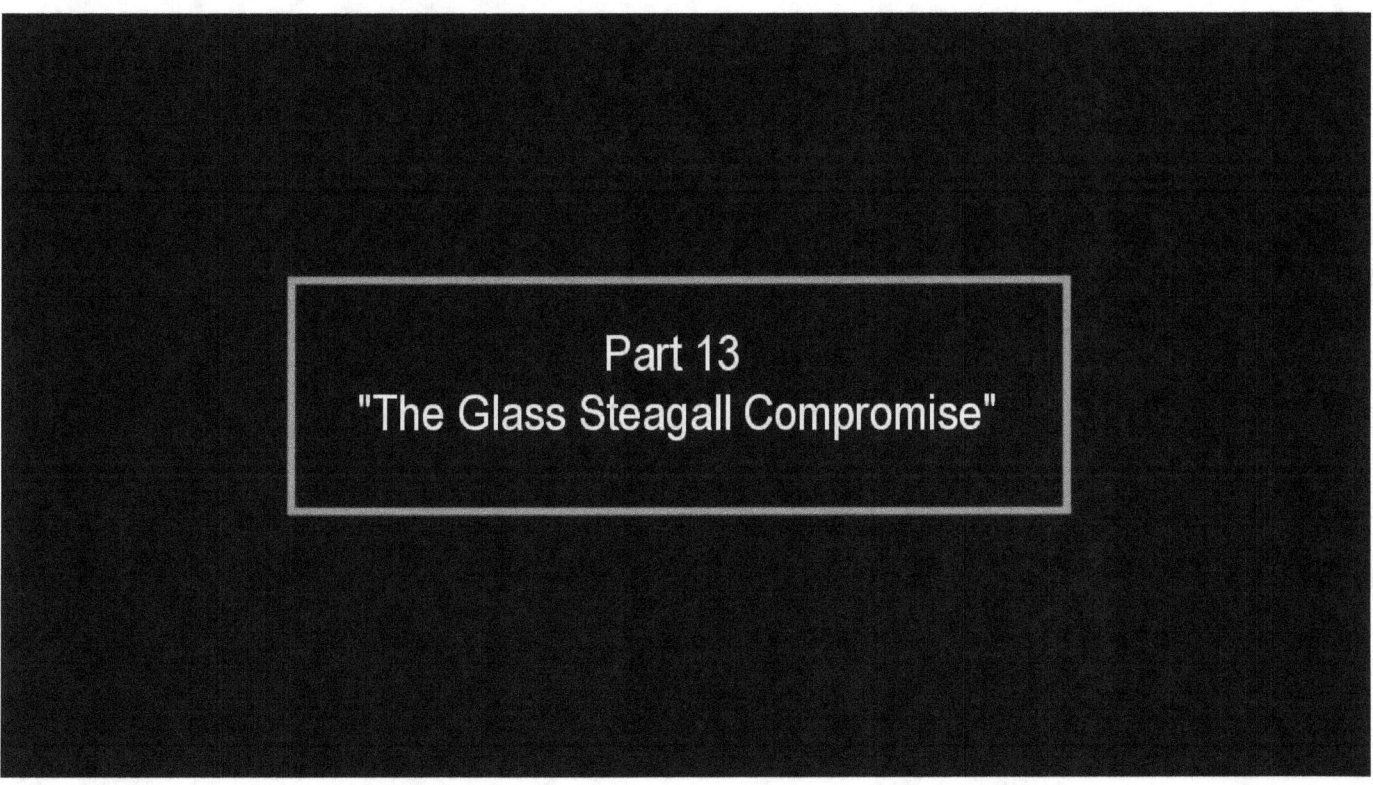

"The Glass Steagall Compromise"

## With the repeal of the Glass Steagall Act

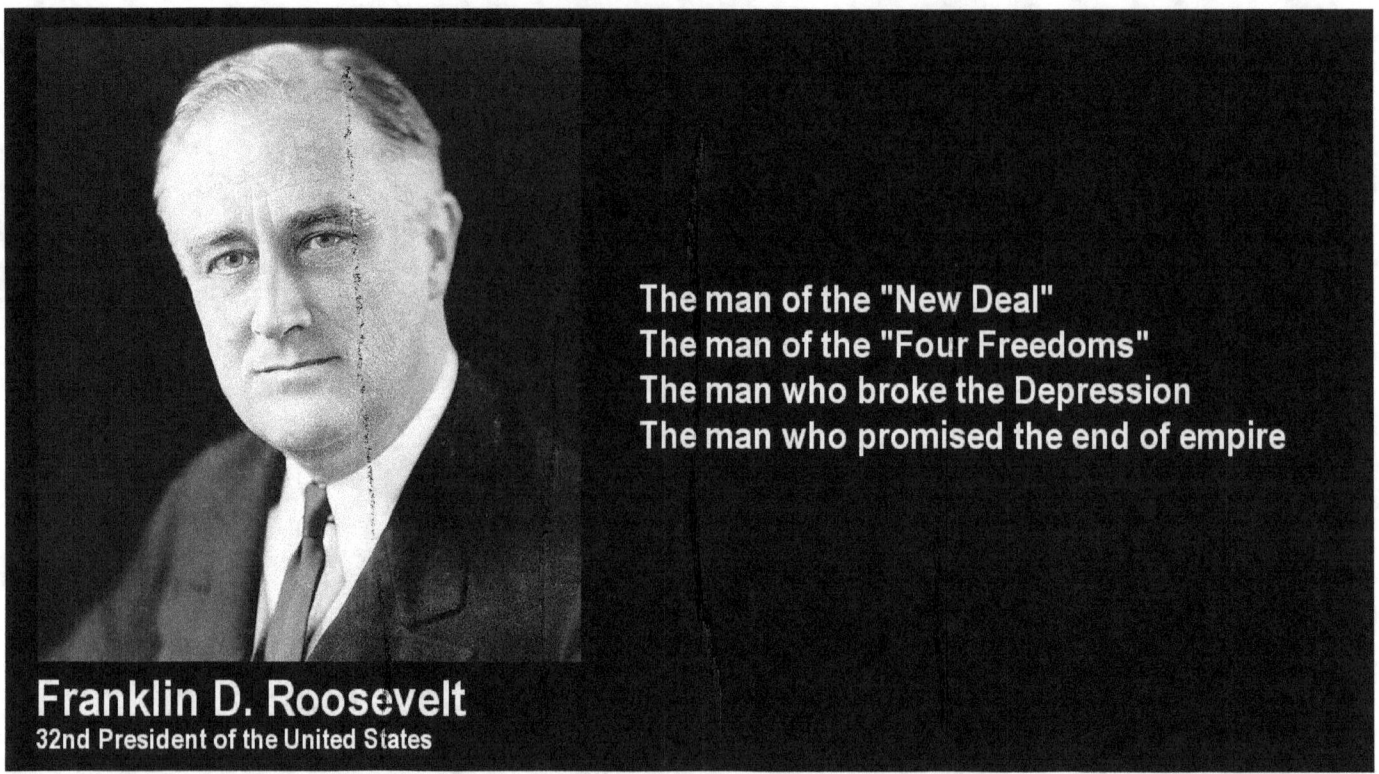

In America the destruction of the nation in the neo-colonial world, was achieved with the repeal of the Glass Steagall Act. The Glass Steagall legislation had been set up during the Franklin Roosevelt Presidency. It was one of the President's first steps to enable America to pull itself out of the economic slum of The Depression that had resulted from the entropy of the looting of society.

# The new colonialism of the Euro empire

In Europe, the same type of destruction of historic economic culture, was achieved with the new colonialism of the Euro empire. The Euro system effectively cancelled all historic structures of the European nations' self-protection and self-development.

In both cases, the grand stealing from the respective nations via various bank-bailout mechanisms, was not only legalized with legislation, but was, in the case of Europe imposed by cleverly arranged treaty obligations, such as the Lisbon Treaty that was designed to be a trap.

## The freedom to steal has become protected

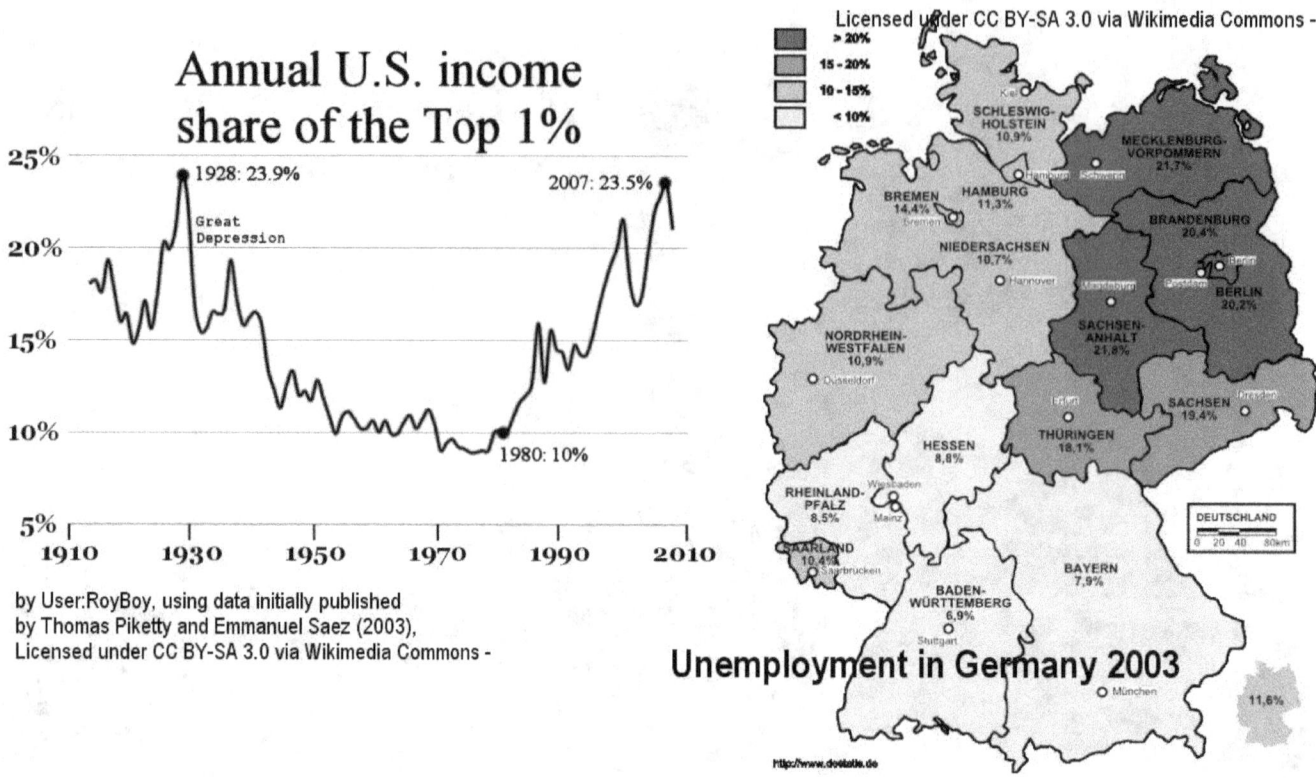

In both areas of the world, Europe and America, the freedom to steal has thereby become protected, while society's self-development has become prohibited.

Both, America and Europe, have so totally bankrupted themselves in this process, as members of the new Flat Earth Society's kingdom of entropy; that preparations are now being made towards the next world war to capture Russia and China as the final remaining resource in the world for continued stealing.

# War against Russia and China

The resulting war against Russia and China, that is now being prepared, will invariably become a nuclear war. All studies have shown this. And the studies have shown too, that nuclear war is unsurvivable.

# In America, the Glass Steagall legislation was repealed in 1999

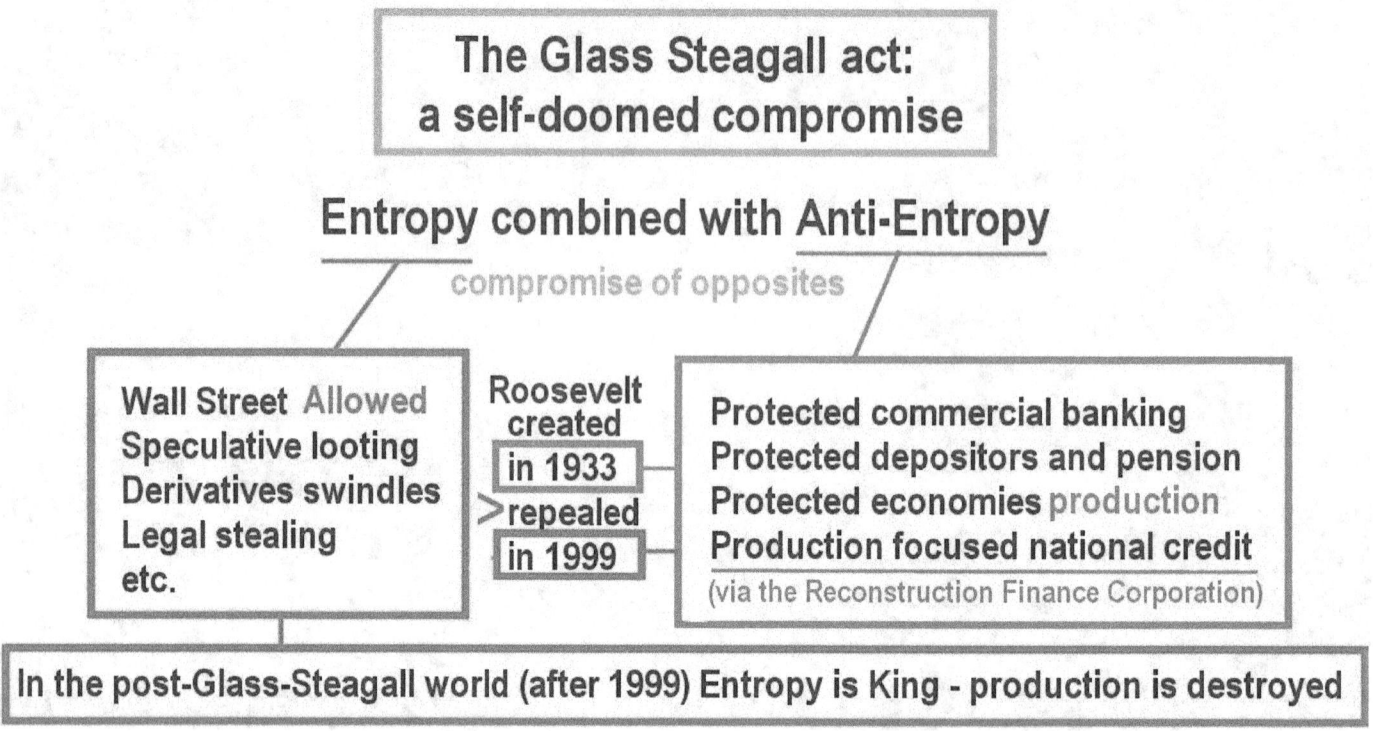

In America, the Glass Steagall legislation had once furnished a type of anti-entropic platform that stood for 67 years and served national development. With it, America had achieved the highest level of general prosperity of any country in the world. It was this foundation for prosperity that was repealed in 1999.

As one might expect, the repeal of the law that had prohibited stealing, opened the flood gates to great national tragedies. Ironically, the resulting tragic failure in civilization was almost assured from the outset. It was assured, because the legislation had been set up as a compromise. The law had compromised, in that it had merely separated the entropy of empire, the kingdom of stealing, from the anti-entropic productive platform that had furnished national self-development and self-protection. It is here, in the fundamental compromise, where its failure is rooted.

## Compromise on principle became its doom

1933 - the Glass-Steagall law to protect that value of the U.S. currency - now repealed

The Glass Steagall law had allowed the entropic platform of the kingdom of stealing, to continue in the background as a compromise. The compromise on principle became its doom. One cannot operate a compromise that incorporates two opposite platforms. The kingdom of stealing named Wall Street, and the principle of national development for Main Street, are irreconcilable. Stealing is destructive. It doesn't create and develop anything.

## Entropy and Anti-entropy are mutually exclusive

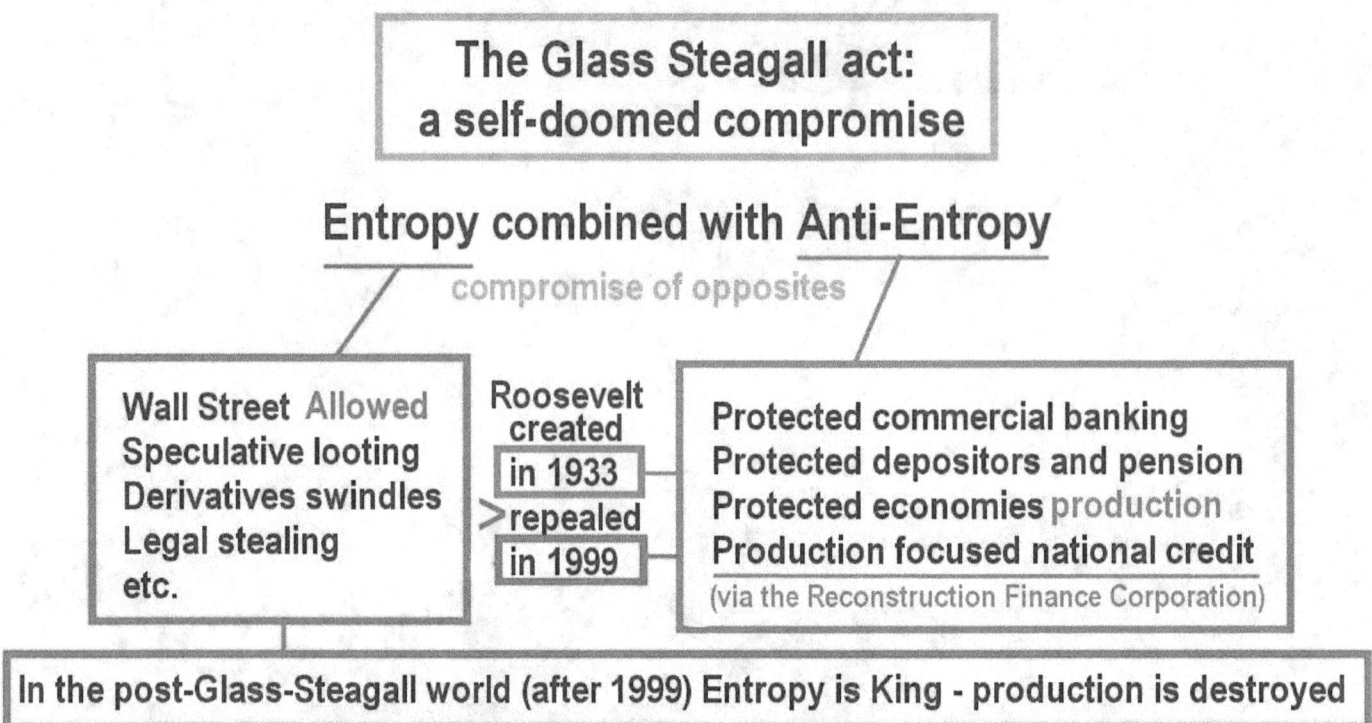

Entropy and Anti-entropy are mutually exclusive platforms. Entropy is not the law of the universe. The universe is Anti-entropic in principle. It is exclusively self-powered and universally self-developing. Diminishment is not a feature of the universe. The universe is forever developing from its infinite, continuing, all-pervading source, and is forever increasing and improving its dimensions in all aspects. Nothing is winding down in the universe. No form of stealing is happening.

## The theory of self-consuming stars

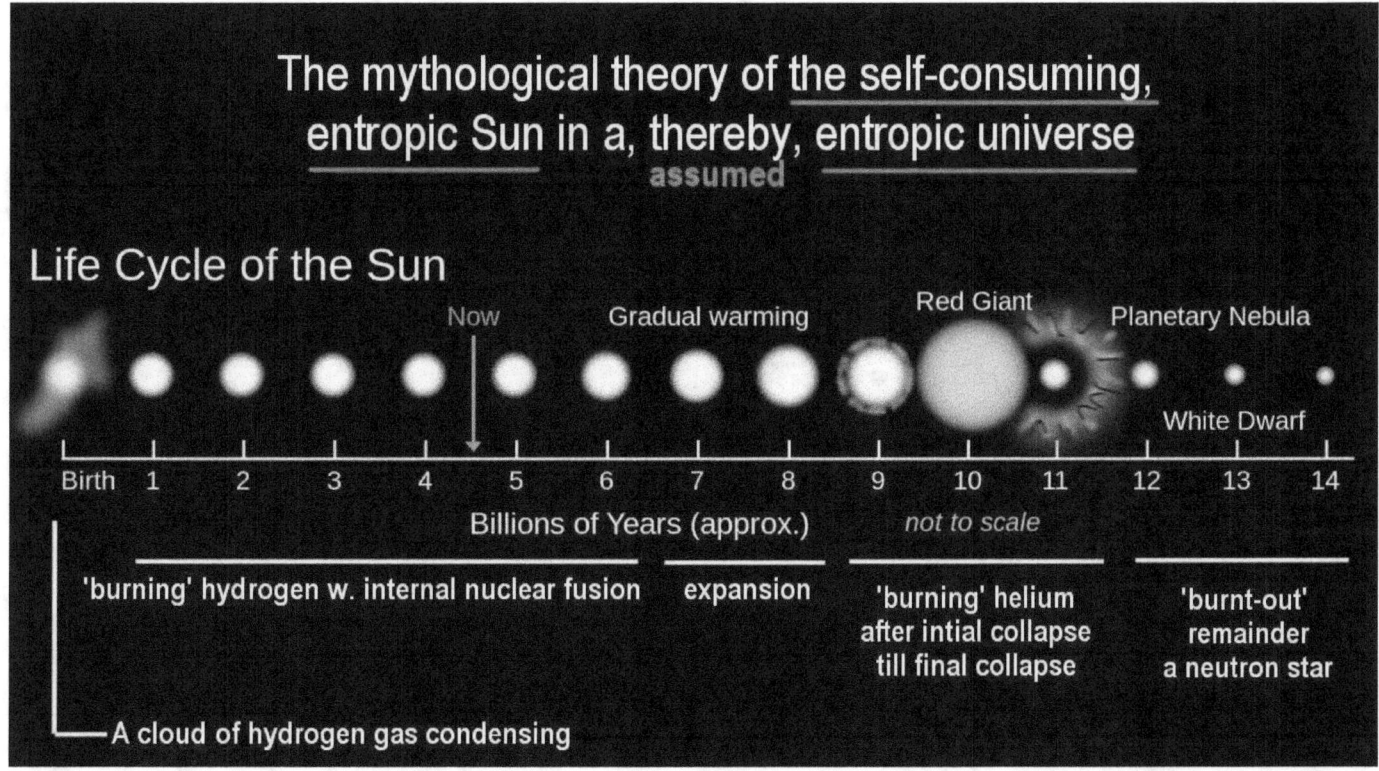

The theory of self-consuming stars that die in the end by their energy depletion, is a myth. The same myth has been falsely applied to humanity and civilization. It is a tragic folly for humanity to reject the anti-entropic platform that the universe operates on, and devise for itself an opposite platform based on a theory that is false in every respect.

The Glass Steagall Act failed because of this cultivated folly of accepting entropy as a quality of natural dynamics. Glass Steagall failed, because the folly of imposing entropy into the premise of civilization had been left unresolved. If the belief in entropy had been overcome, empire would have been closed down, and the development of the USA would have continued.

But by its compromising on a fundamental principle, the Glass Steagall Act has stood effectively in self-denial from its very beginning. This self-denial eventually opened the door to its doom.

# We play the same compromising game

We play the same compromising game by allowing nuclear war to stand in the world as an option for warfare. It is an element of the entropic kingdom of stealing, an element that has remained unresolved. This element cannot be resolved by itself in isolation. The ever present nuclear war danger can only be resolved by scrapping the entire package of empire that it is a part of. This means stepping up to the anti-entropic platform for civilization, which is the natural platform for humanity.

## The Glass Steagall compromise

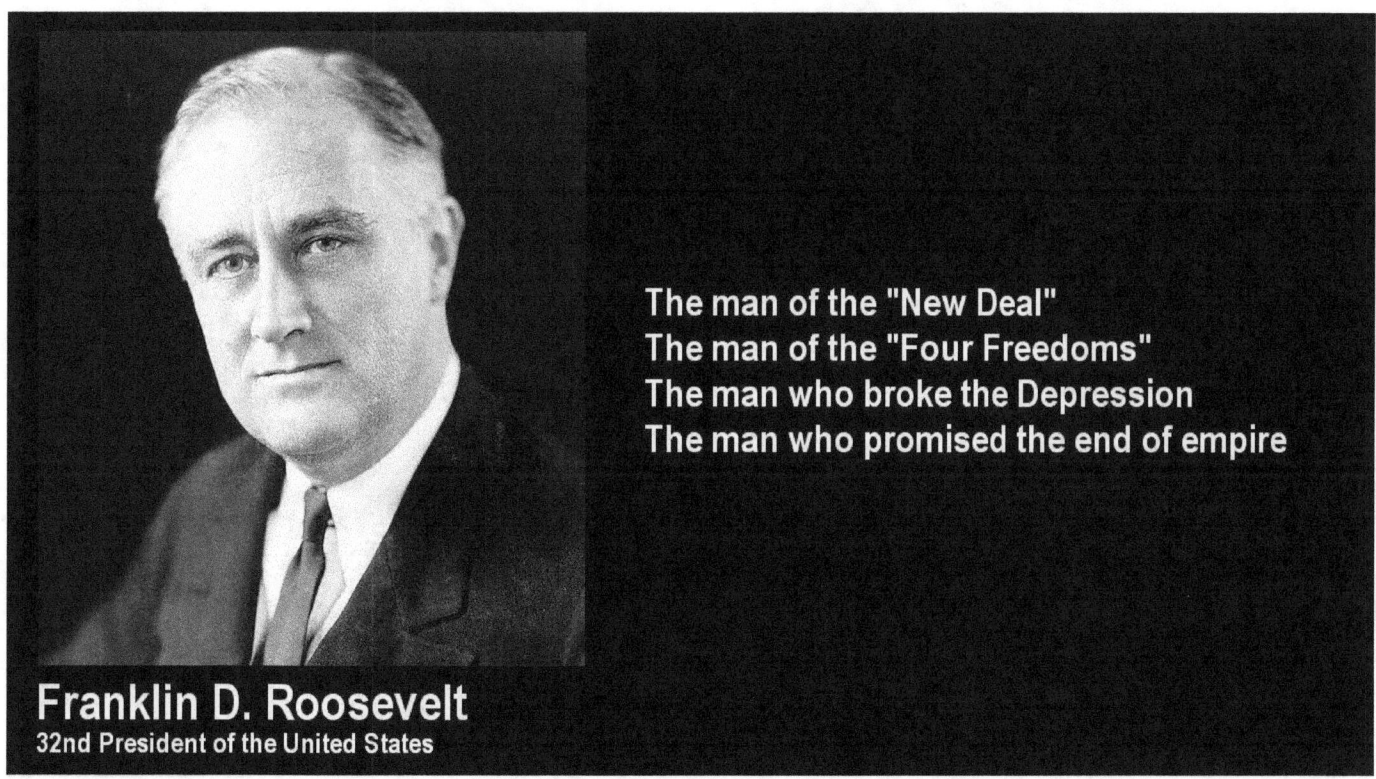

It appears that the Glass Steagall compromise was not made entirely by choice in 1933. It was allowed by default, because the new President, President Roosevelt, with all the popularity that he had, with which he had secured his election, couldn't muster the political backing in the house of government to eliminate the entropic platform of financial thievery from the national landscape. The compromise came from that. In like manner, the political backing is still far out of sight to banish nuclear war that looms as the final stage of political entropy.

## Twenty years before Roosevelt

By the foolish compromise that was allowed in 1933, the kingdom or entropy had survived and become a monster. The monster should have been cast out right then. Instead it was permitted to continue as the private owner of the financial house of the USA in the form of the Federal Reserve system.

The tragedy had it origin long before Franklin Roosevelt's time. Twenty years before Roosevelt the entropic kingdom of thievery had already corrupted the nation so deeply, that it was able, with corrupted politicians, to steal the nation's currency into private hands, by an operation that became deceptively named a "Federal" system. From this lavish base of complete control over the nation's money, which the Glass Steagall Act had continued, the nation's leaders were subsequently further corrupted to repeal the Glass Seagall Act - the very platform on which the nation's prosperity had been built, in order to get the nation back to the Depression that the Federal Reserve system has created shortly after it was formed.

This immense corruption of the U.S. Congress and Senate, to repeal Glass Steagall, was achieved in 1999 with a giant slush fund from the kingdom of money amounting to hundreds of millions of dollars. These huge sums, handed out under the table to a selected few, did buy the terminating votes from the small-minded who are easily turned to become traitors for hire.

## The year 1999 marks the historic beginning

### Since 1999

President Bush elected (the war and terror President)
The September 11, 2001 State terrorism event
'Perpetual' War against Afghanistan (2001 till end of 2014)
Iraq War (2003-2011)
Legalizing of Torture
Collapse of the auto industry
Home foreclosure crisis
War threats against Iran
President Obama elected (the shutdown President)
Great bailout Bank Heist ($50 trillion stolen since 2008)
Healthcare and social security decimation
Libya color revolution (war) - (2011 - to the present)
Egypt color revolution (war) - (2003 - to 2004)
Syria color revolution (war) - (2012 - to the present)
Ukraine violent overthrow as a step towards Russia
Nuclear War threats against Russia and China
with the western financial system dead on its knees.

...we've stood closer to nuclear war than ever before, and more often.

### The Kingdom of Entropy the post-Glass-Steagall era
1999 to the present
**an era of Extreme Stealing Terror, and Perpetual War.**

Annihilation is assured

500,000 times Hiroshima in one hour

Castle Bravo - the first U.S. test of a dry fuel thermonuclear hydrogen bomb - March 1, 1954 at Bikini Atoll, Marshall Islands

The rest is history, as people say. This history is still unfolding. The year 1999 marks the historic beginning of what may be called one day, the greatest national tragedy in the entire existence of the USA.

The tragedy started with collapsing financial values that prompted the infamous 911 State terrorist event, which in turn prompted the 'Perpetual' War doctrine, supposedly to fight terrorism. On this track war was brought to Afghanistan, then Iraq, with the legalization of torture along the way. At home in the USA, the collapse of the auto industry decimated employment, with war threats against Iran occurring in the background, while the home foreclosure crisis unleashed social chaos.

Since this didn't solve anything on the financial front, the Greatest Bank Heist in the history of the world was staged seven years later, this time not to rob the banks, but to rob society of upwards to $50 trillion to bail out the gambling casinos that the banks had become. Of course, to save money, society's healthcare and social security system was decimated, which is still ongoing.

Since this immense sacrifice didn't solve the financial collapse crisis either, the scourge of war was brought against Libya to murder its leader in the name of liberty, to liberate its oil resources. The liberty-revolution, that was deemed a splendid success, became the blue-print to destabilize Egypt and later Syria, to depose their governments likewise. And since none of this helped to slow the financial collapse, the elected government of the Ukraine was violently overthrown by lavishly financed 'hired' Nazi 'revolutionaries' as a stage for war against Russia, and by alliance, also China. Any war against Russia, invariably becomes nuclear war that involves both, Russia and China, which adds up to a madness that is unsurvivable.

That's where we stand today, with the western financial system now almost totally dead on its knees, which not the greatest sacrifice in the world can revive.

# To restore the Glass Steagall law

1933 - the Glass-Steagall law to protect that value of the U.S. currency - now repealed

A substantial political movement has begun in recent years to restore the Glass Steagall law in order to halt the train to hell.

Obviously, if the law had not been repealed, the economic, financial, and strategic tragedy that the nation of the USA, and also the world, has suffered, would likely have been prevented. However, reinstating the Glass Steagall law at the present stage is no longer sufficient. Too much has been destroyed on many fronts that storing an old compromise would solve the crisis that has become a national tragedy.

## Solving the tragedy at this stage

Solving the tragedy at this stage would require an uncompromised stand for the principle of anti-entropy being reflected in the universal self-development of humanity on the entire front of civilization. Nothing less would do. This would include anti-entropic finance and economics, large-scale scientific and technological development, national banking, national directed credit creation, quality education, health care, universal high-quality housing, and a commitment to uplifting culture. This complete shift to an uncompromised, anti-entropic standpoint on the entire front, would leave no room for stealing.

# To merely reinstate Glass Steagall, defies the nature of reason

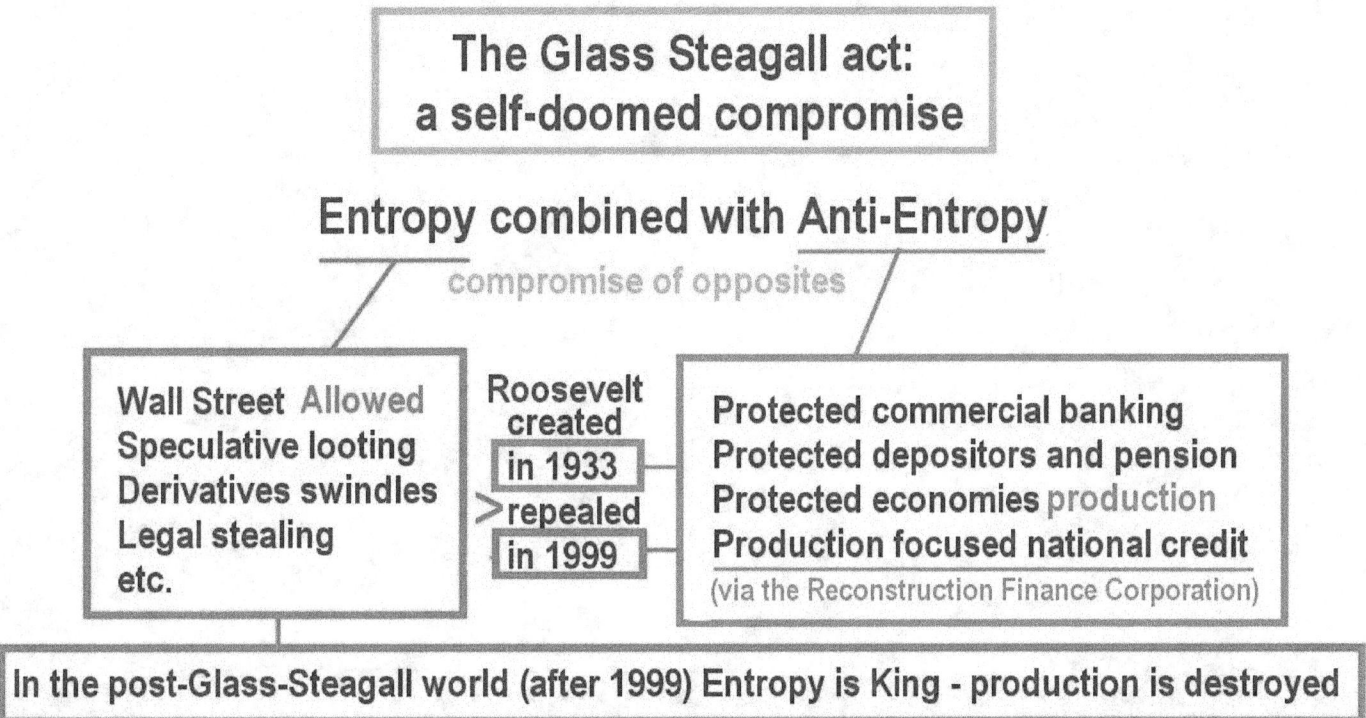

The suggestion to merely reinstate Glass Steagall, defies the nature of reason, which is itself anti-entropic. For example, why would we restore elements of a failure that has collapsed the system itself, which the kingdom of entropy brought about, and which was tolerated under Glass Steagall? Or why would one bring back even the anti-entropic element of Glass Steagall that had been insufficient in itself to eclipse the entropic elements. It appears that the Glass Steagall compromise was made, because the principle of anti-entropy had not been developed extensively enough to be understood in 1933.

## Meeting the Ice Age Challenge

This sets the stage for the surprising recognition that the platform for meeting today's vastly larger challenge, the Ice Age Challenge, must rest on a dramatically higher level than just getting back to an old bill that had failed.

## To create a more-just economic order in America

1933 - the Glass-Steagall law to protect that value of the U.S. currency - now repealed

Glass Steagall had originally been designed to create a more-just economic order in America, in which society is protected against the ravishing rule of entropy. While this law has met the most pressing needs of the time and has enabled unprecedented prosperity and stability, it falls short of meeting today's requirement for meeting the Ice Age Challenge where vastly greater imperatives stand before us than the 'little' concerns that the Glass Steagall Act had dealt with.

## Bring the future demands into the present

In order to meet the great challenge that is no longer avoidable, the self-rescue of society needs to begin with the principle of anti-entropy, which needs to overturn all previous laws, falsely written, that don't measure up to its standard. Humanity needs to do this all over the world, to be true to itself. Humanity, is the only anti-entropic species on Earth that has the capacity with its intellect to look at the universe, discover its principles, and with the discoveries leap ahead of illusions and notions and reach deep into the future and bring the future demands into the present for the shaping of policies in order that life can be preserved, and a rich civilization arise, three decades from the present in a radically altered world that no one in remembered time has ever seen or experienced, but which can be known in the mind.

# Segment 5 - The Need for Looking Forward

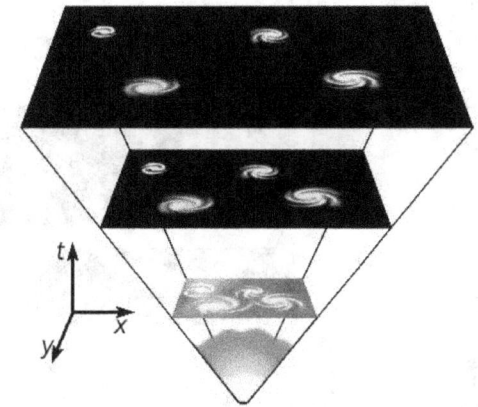

Big Bang Blow-Out

Segment 5 -    The Need for Looking Forward

By remaining latched to the past in science, economics, and politics, humanity stands in denial of its anti-entropic nature, and its thereby infinite potential that is already demonstrated in principle by humanity's amazingly expanding population by means of scientific, cultural, and technological progress.

The amazing evidence stands as a beacon for us in the present, for meeting the vastly greater requirements on all of these fronts for meeting the Ice Age Challenge. This includes not only anti-entropic energy and freshwater development on a worldwide scale, but also anti-entropic economics to get the needed infrastructures built before the already-near Ice Age Transition occurs in potentially the 2050s.

This gigantic scene of opportunities takes us far beyond the entropy of the Big Bang theory and its empty center. It takes us all the way to rediscovering ourselves as the supreme being on the Earth that has just begun to peck open its shell. On this basis the California Water Crisis con be solved, such as with the principle of Deep Ocean Reverse Osmosis Desalination, which is completely self-powering, causing rivers of freshwater to flow out of the ocean, independent of climate.

## Nothing is gained from clinging to the past

Restoring an old platform from the past that has failed, whether it is political in nature like America's Glass Steagall banking law that had enabled America to become the most prosperous nation in the world, but which has failed, or whether the platform from the past is scientific in nature, such as the more obviously failed Big Bang theory, any such clinging to the past is ultimately an exercise of futility that in effect celebrates entropy, the very sense of entropy that the false Big Bang theory places before humanity. Nothing is gained from clinging to the past, as inviting as this may seem, because to do so denies our power to live in the future, where we can look forward to greater achievements than any that have been wrought in the past.

# By becoming latched to the past

By becoming latched to the past, people live by the entropic assumption that humanity doesn't have the capacity anymore to leap ahead into the future and pull the present up behind it.

## By staying latched to the past

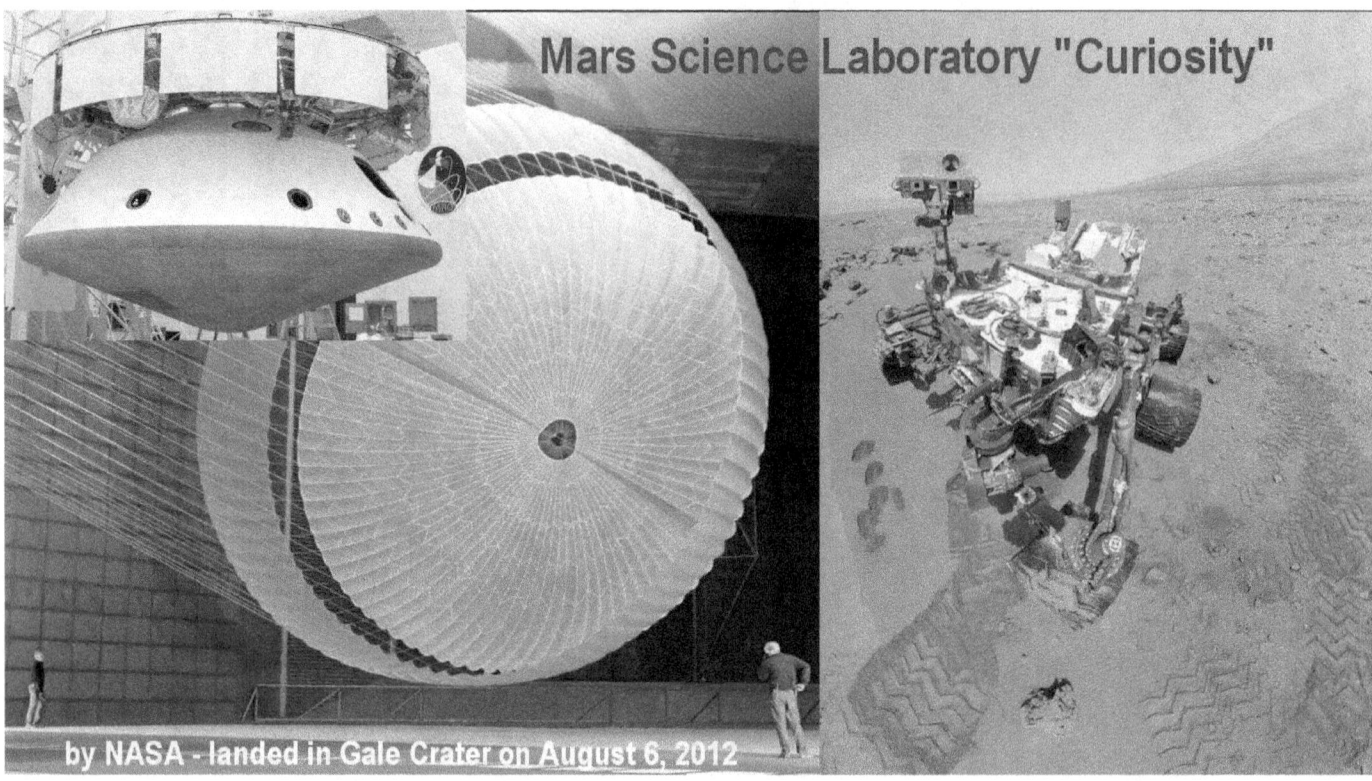

By staying latched to the past, people lie to themselves. They lie to themselves about their own humanity, because it is knowable that each person as a human being has the capacity to know the truth. This applies in science, economics, politics and in civilization as a whole

# Without the advancing recognition in society

Mang Lev
431 km/h
top service speed

Pudong International Airport
Shanghai
China

Without the advancing recognition in society of the anti-entropic nature of the human being, which no other forms of life on earth can match, humanity would doom itself, both in the present and in the future, because our very existence depends on our human, ever-expanding, creative capability, and increasing scientific and technological progression.

## Glass Steagall has become too shallow

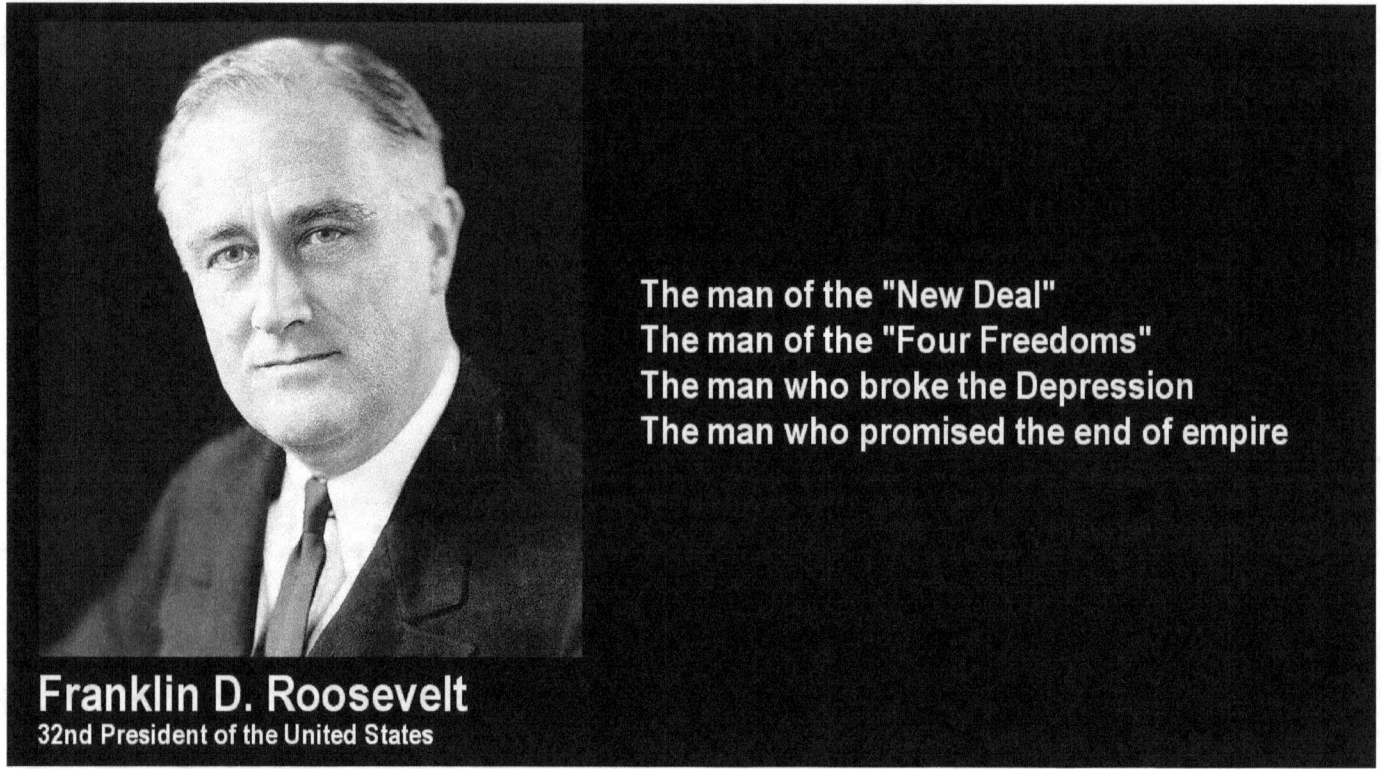

This means, that the imperative that the Glass Steagall act in 1933 had been based on, on the political front to advance America's economic progress that was much needed 65 years ago, has become too shallow to make the grade in today's world. We need to go further, by a long way. Compromises, no matter how tempting they may be, are no longer useful, nor beneficial. Any compromising on the grand scale of civilization has become too dangerous for the whole of humanity, because the imperative in today's world is that we move ahead on all fronts without fail, because the Ice Age Challenge cannot be met with anything less.

# The imperative comes with the Ice Age Challenge

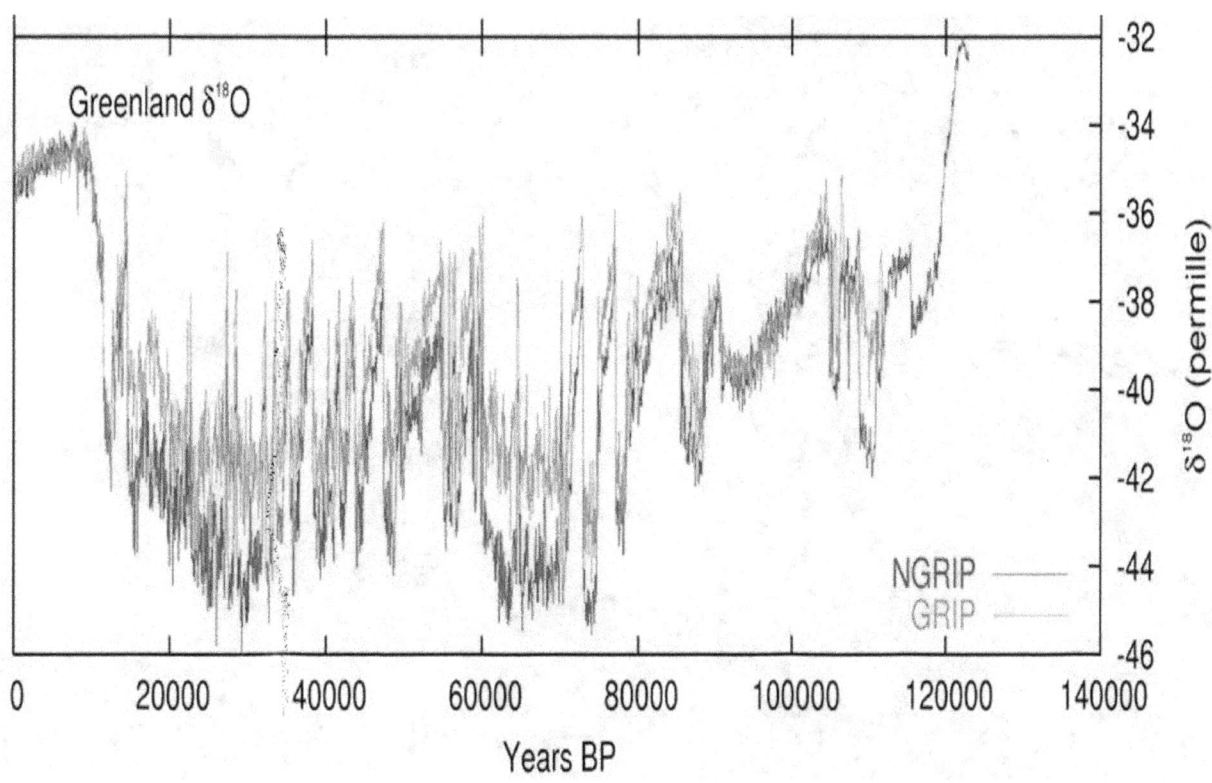

The imperative that inspires us to become more human, comes with the Ice Age Challenge that cannot be side-stepped or be compromised with. Nothing less will do, than moving ahead as effectively as possible.

## It challenges us all to become human beings

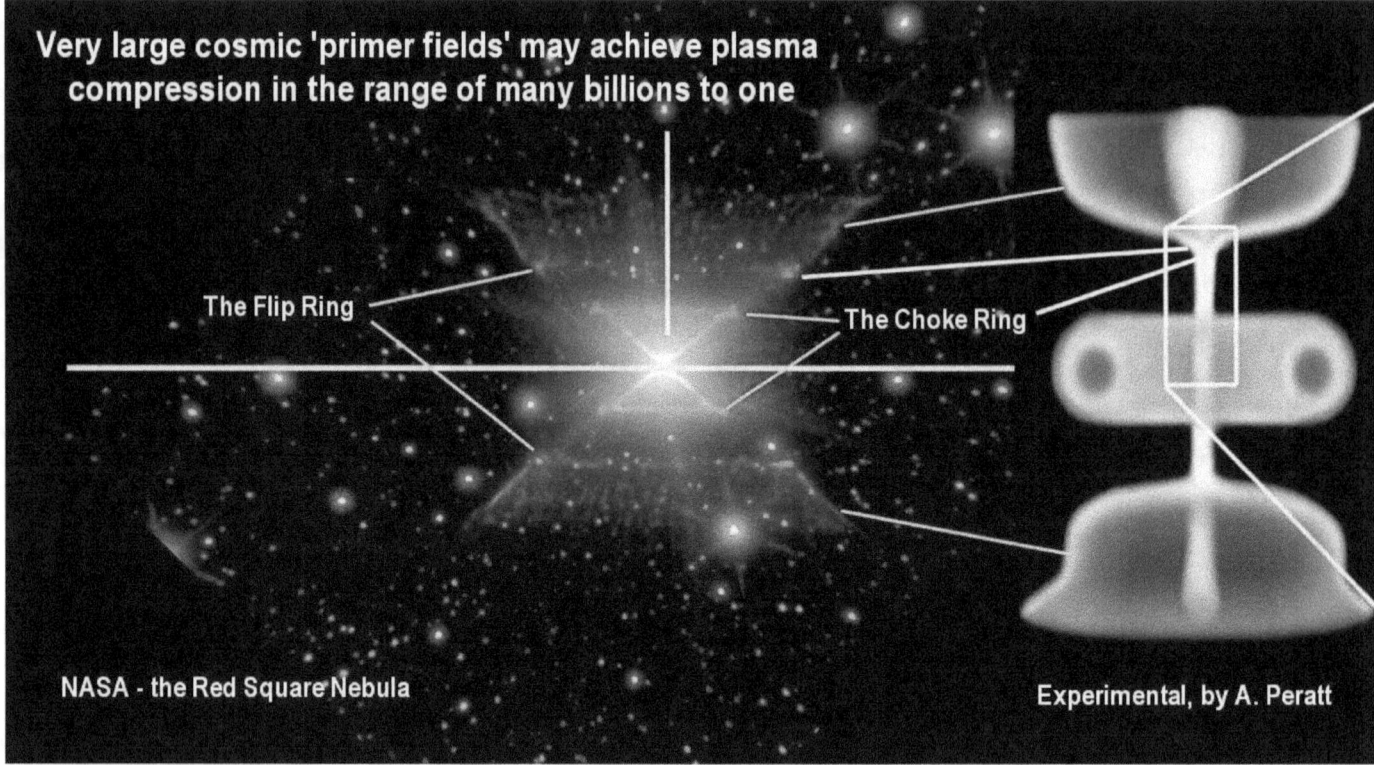

It challenges us all to become human beings in the fullest sense of our highest self-recognition, and that we move with the universe by utilizing its resources, and the resources we have within as children of the universe, who have the power to become one with the universe. The demand, that we place on us, takes us far beyond the falsely imagined Big Bang entropy.

## To create an Ice Age Renaissance

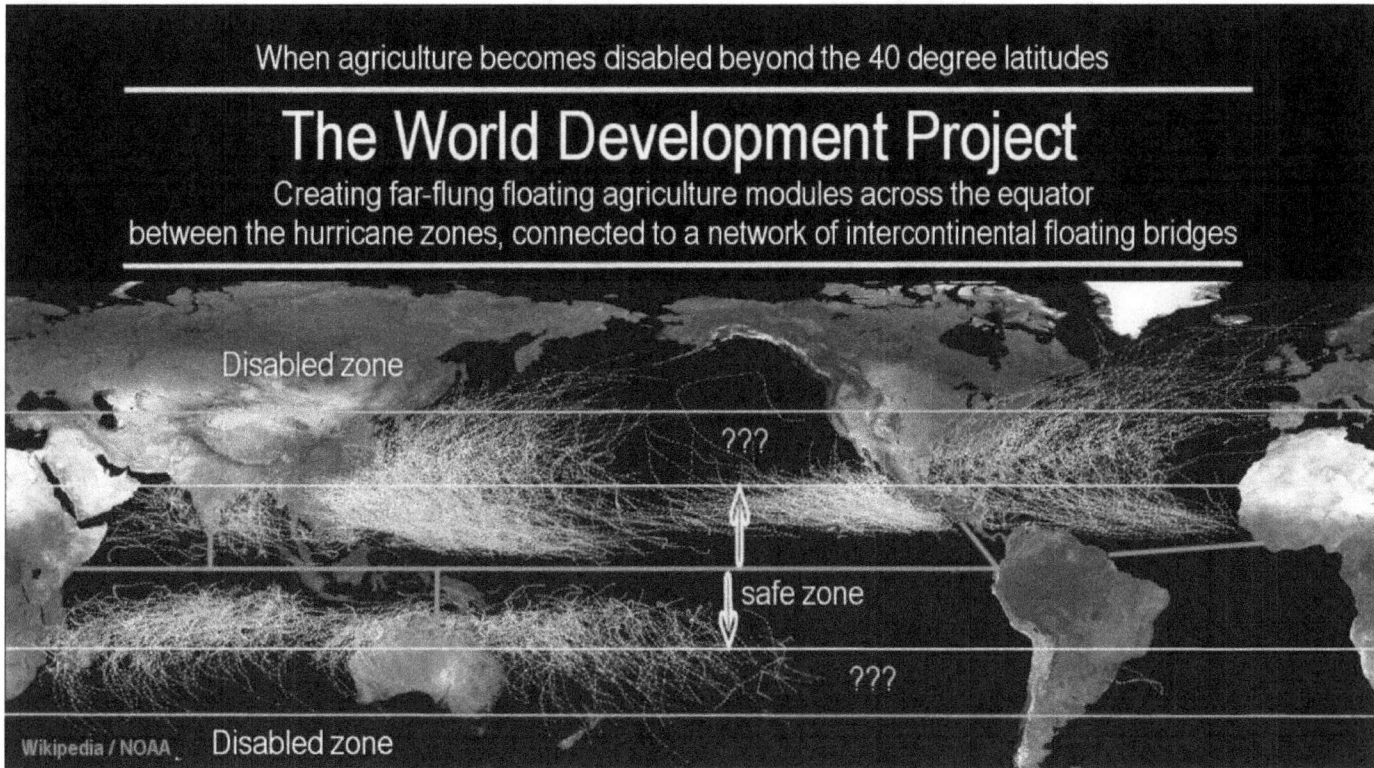

We place this demand on us, for the obvious reason that the Ice Age Challenge can not be met with less than the complete commitment in society to the principle of universal anti-entropy that is reflected in humanity worldwide, just as it is prominently apparent in the operation of the universe, and on earth in every form of renaissance that ever was. The dynamics of the universe challenge us, to create an Ice Age Renaissance for ourselves, on a scale that far supersedes our grandest dreams to date, because we have become accustomed for far too long, to dream too small.

# We need to look forward with the eyes of science

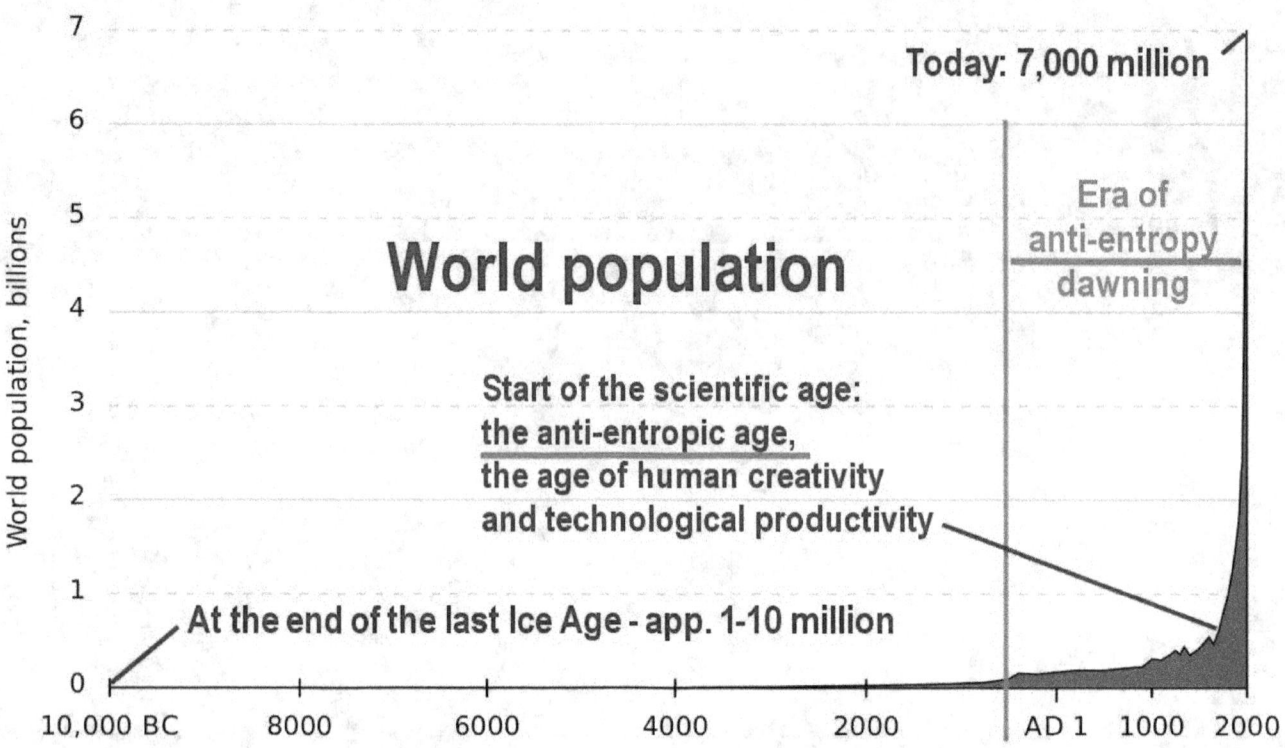

The point is that we need to look forward with the eyes of science, and consistently step beyond the platforms of the past and their limits and failures, even as we built on past achievements and experiences.

## In terms of our natural capacity as human beings

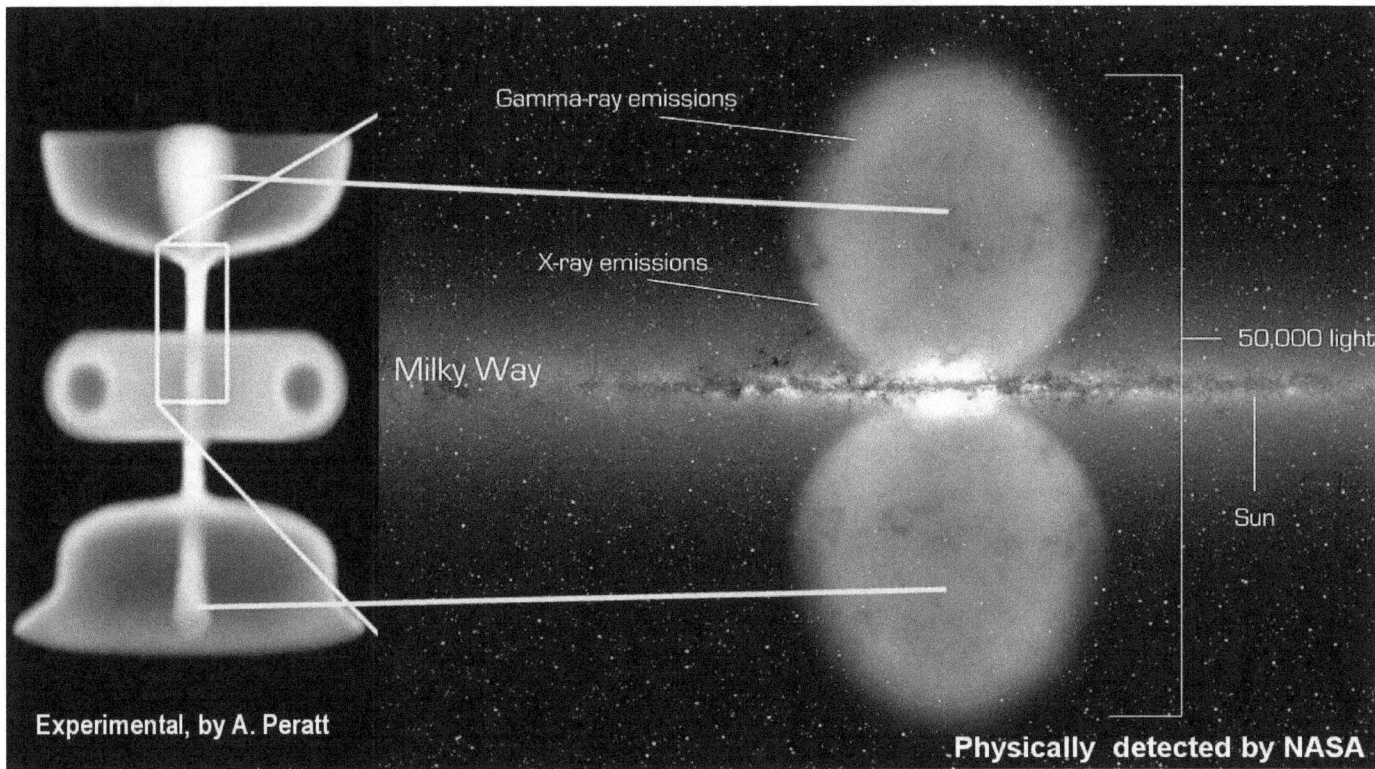

This means that we focus on what is real in the universe and in ourselves as human beings as a part of the universe, and that we use this higher-level focus as a starting point.

In terms of our natural capacity as human beings, we are presently operating at a large deficit on the road towards meeting the Ice Age Challenge. Too many centuries have been wasted under the chokehold of entropy in the kingdom of thievery, such as the Roman Empire and the like, have placed on us.

# The Ice Age phenomenon is not the product of entropy

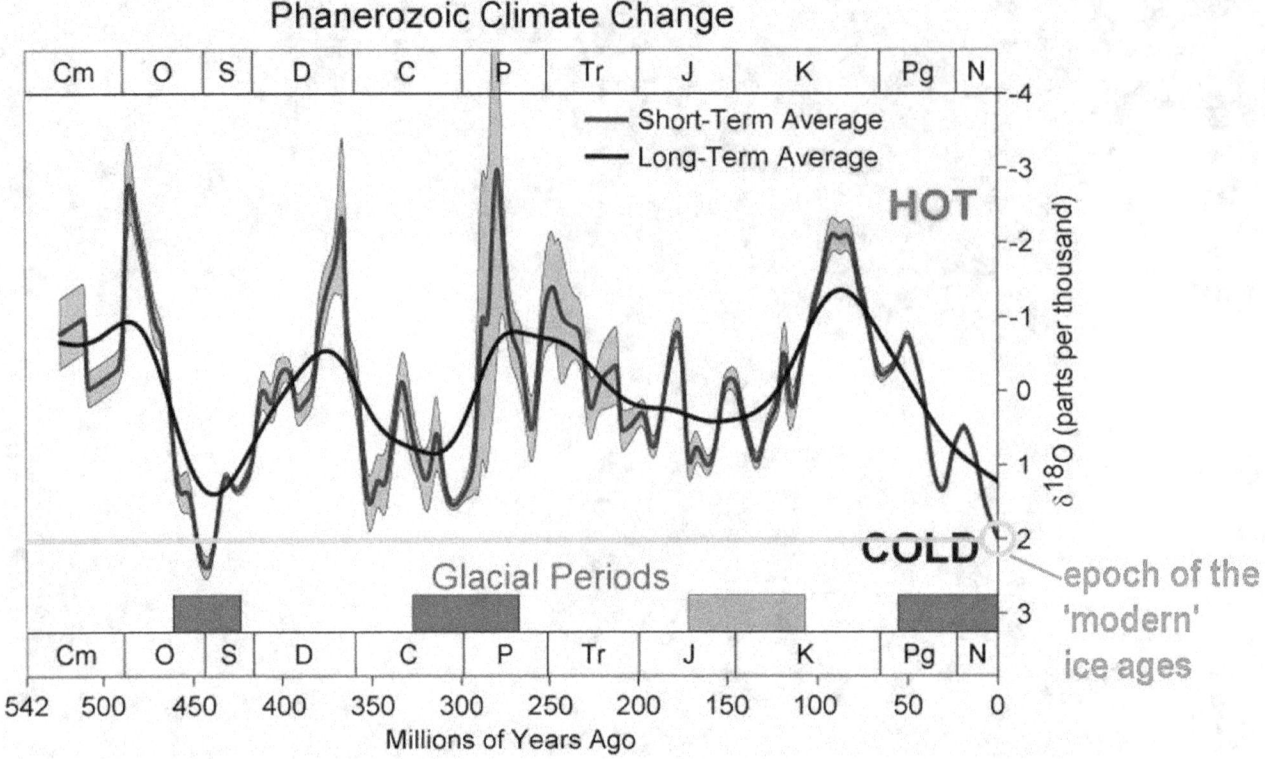

The Ice Age phenomenon by itself, is not the product of entropy. It is a phase of the anti-entropic cycles of the universe, and of ever-higher forms of life unfolding within them.

# Children of the anti-entropy of the universe

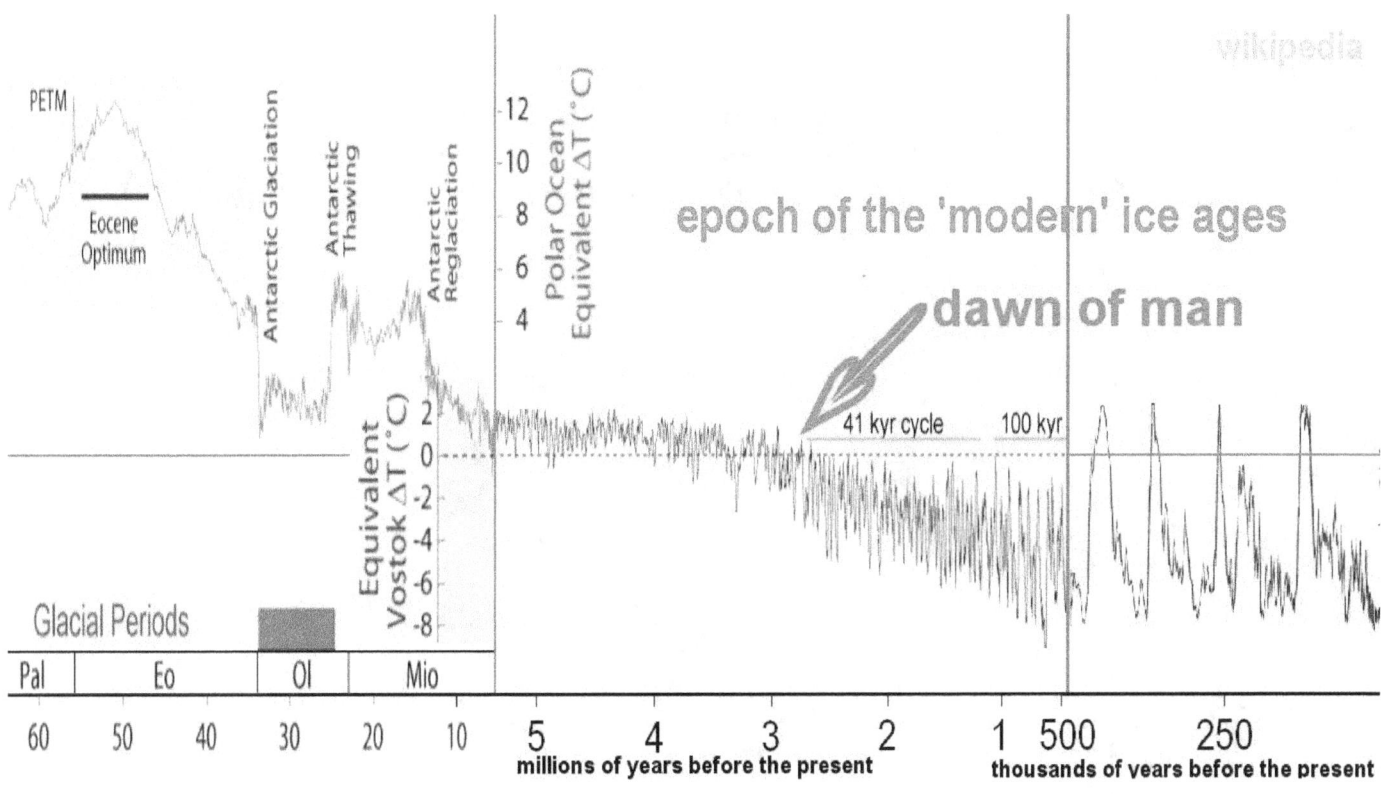

We, ourselves, as intelligent human beings, are the children of the anti-entropy of the universe that unfolds evermore forms of beauty, sublimity, and power, almost in its own image.

It is highly likely that the dawn of humanity on Earth had not been possible on earth until specific astrophysical conditions had existed that had enabled a breakthrough in the development of a higher-level species, and by several successive breakthroughs that have occurred subsequently, which would render us the children of the universe.

We may fear the Ice Age, but in real terms the ice ages have been our cradle in which substantially larger volume of galactic cosmic-ray flux have reached the Earth for most of its time, which didn't happen before our time.

We, as humanity, emerged from the cradle of the ice ages as the brightest stars in the heavens of life, and with a potential direct connection with the universe itself, a connection that we have just begun to discover.

According to all evidence, we are on a track that takes us far beyond the notion of entropy. We are not glorified animals, we never have been, but are the diamonds in the heavens of life.

# Ice ages are critical elements in the progressive dynamics

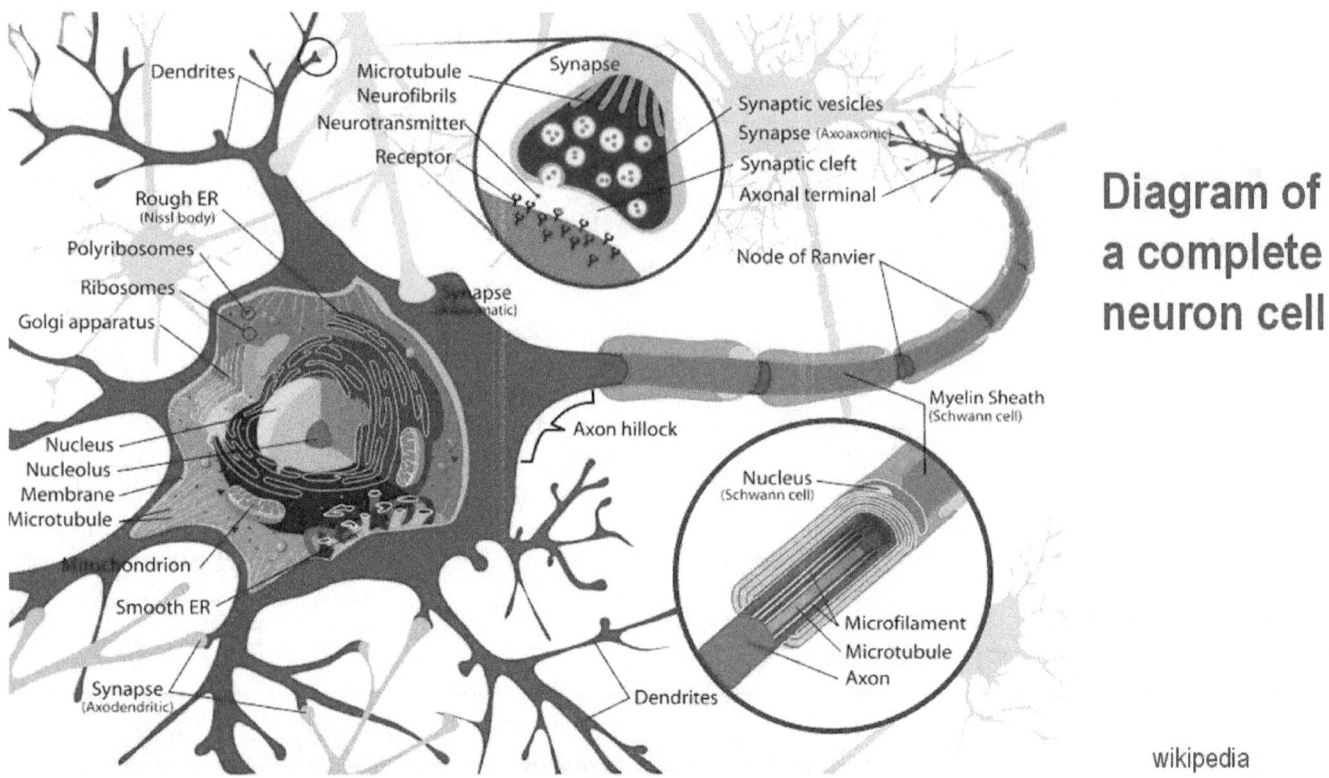

Diagram of a complete neuron cell

wikipedia

The cosmic-ray flux that has affected us richly through our past, is an electric phenomenon, a phenomenon of fast moving electric particles, and related particles. This fast electric flux has the potential to be a critical factor for the development of complex neurological systems as we have within us that have become systems of cognition, and ultimately, consciousness. In this context we may be the children of the creative, anti-entropic, quality of the universe in the most intimate sense, where the ice ages are critical elements in the progressive dynamics.

# The Big Bang theory stands plainly in denial

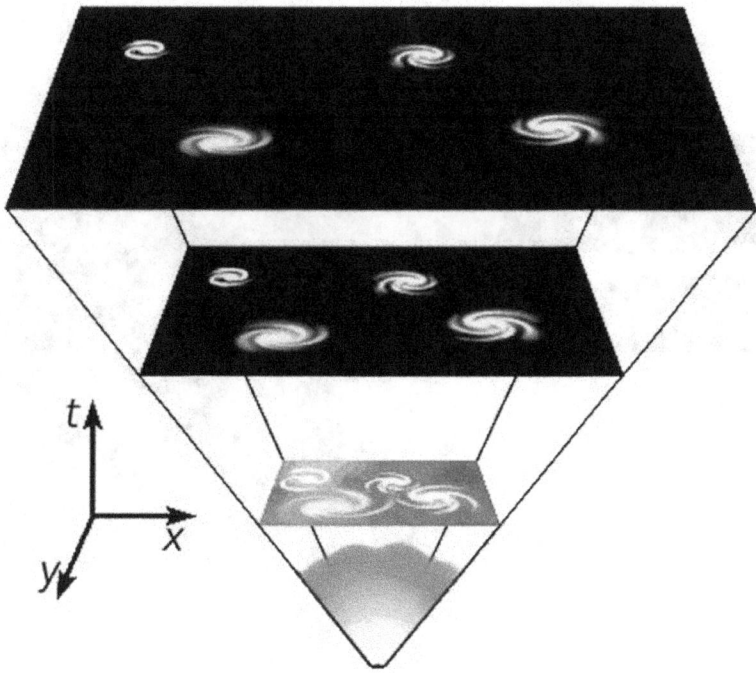

The Big Bang theory stands far, far distant from the anti-entropic principle of the universe, though it took us a long time to discover the principle that is expressed in our humanity.

The discovery that we have made on so many fronts, tells us that the Big Bang theory stands plainly in denial of what is self-evidently true.

# Big Bang theory might have been intentionally staged

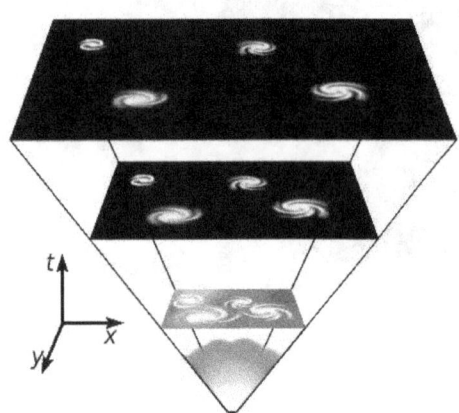
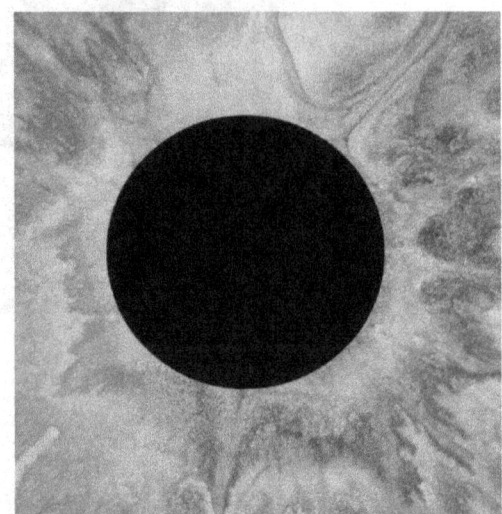

While no impelling evidence exists that the astrophysical Big Bang theory had been intentionally developed for political purposes, and might have been promoted as a cultural warfare project to create an empty center in society - in science, economics, and in culture - the timing of the promotion of the theory - as a counter-ideology - seems to suggest that the Big Bang theory might have been intentionally staged for such intentions.

## The Big Bang Cosmology

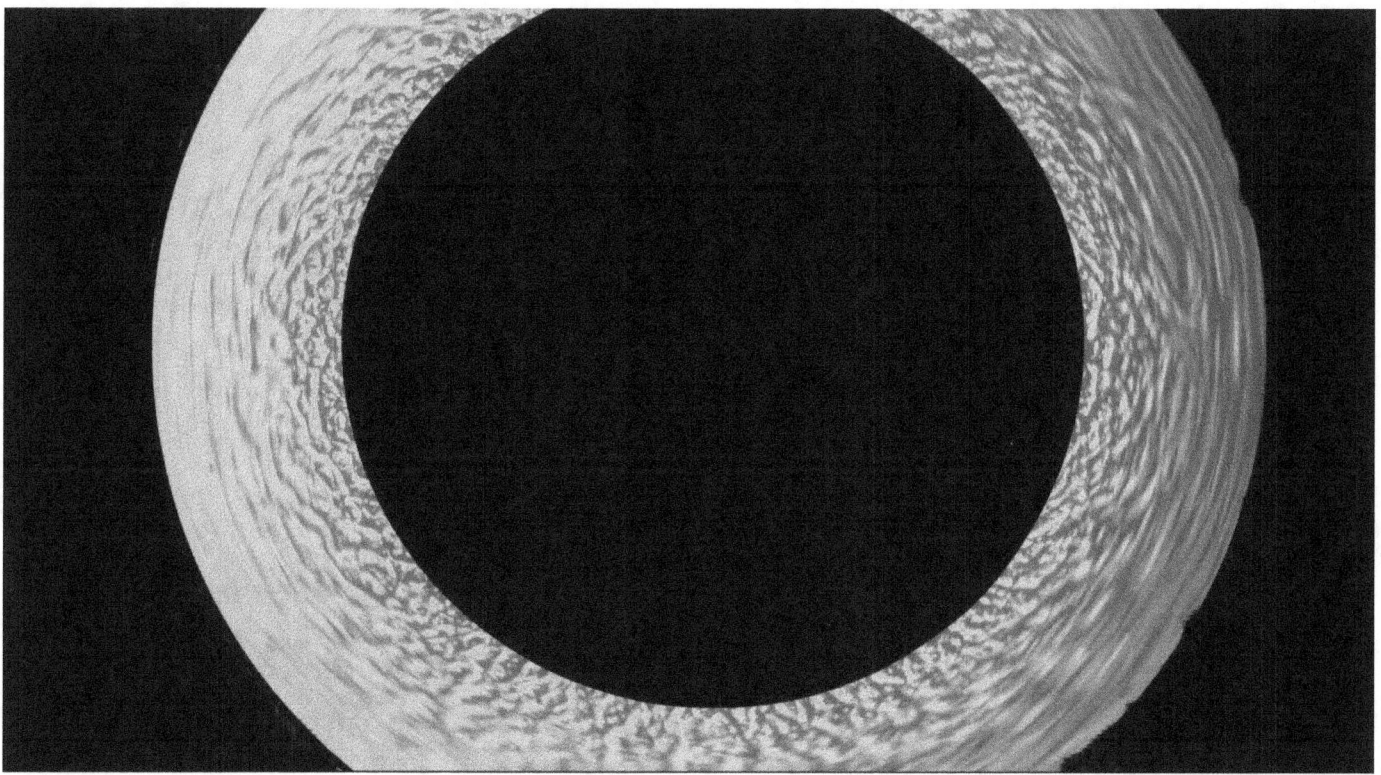

This factor, all by itself, should inspire us to regard the Big Bang Cosmology and its fire of entropy with an empty center, as nothing more than a tragic product of false assumptions at best, and false intentions at worst.

## The Ice Age is near, as near as the 2050s

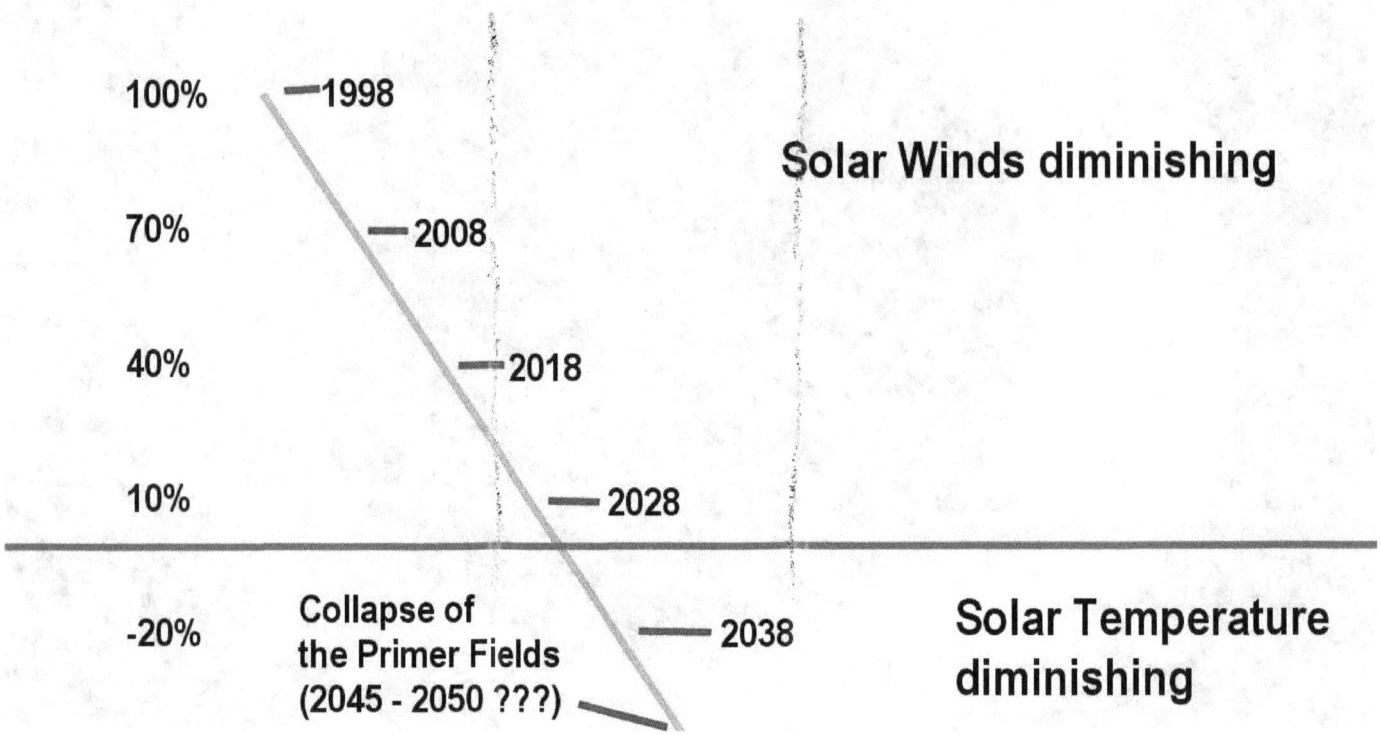

The Ice Age is near, as near as the 2050s, potentially. With the challenge that this poses, in mind, it shouldn't be too hard for society to step up to higher ground and move forward from there to build itself a correspondingly great renaissance, on the basis of this truth, with which to rebuild its deeply-collapsed civilization in order that the Ice Age Challenge be met in time, before the physical Ice Age starts anew.

# A great need for a renaissance of truth

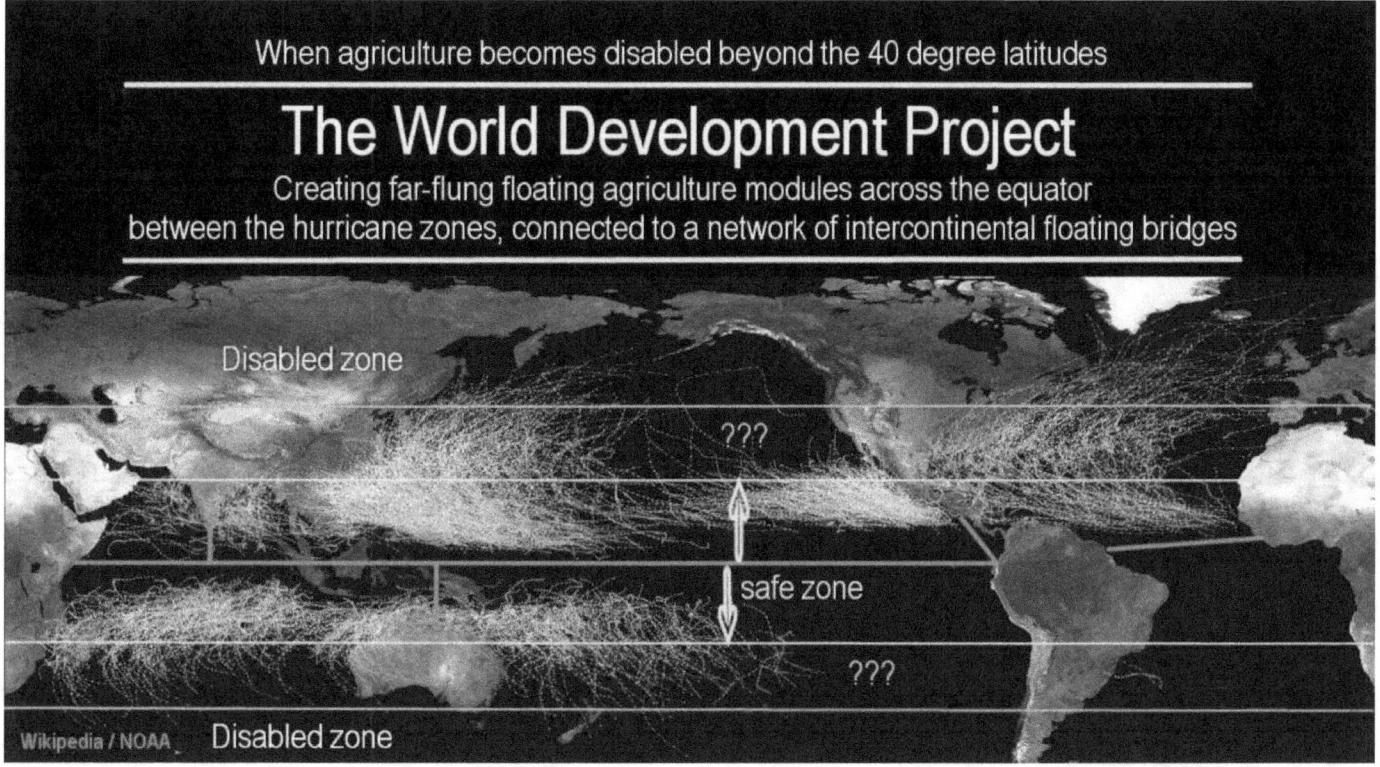

We have a great need for a renaissance of truth towards this end, because the truth is, after all, the builder of worlds, and truth does not diminish.

The truth is here to stay, and the truth is, that humanity is an anti-entropic power in the world. For this, we have ample evidence. In truth we find the foundation for our future. "In Truth We Trust." That's what the new banners proclaim.

## The proof of the 'pudding' is unmistakable

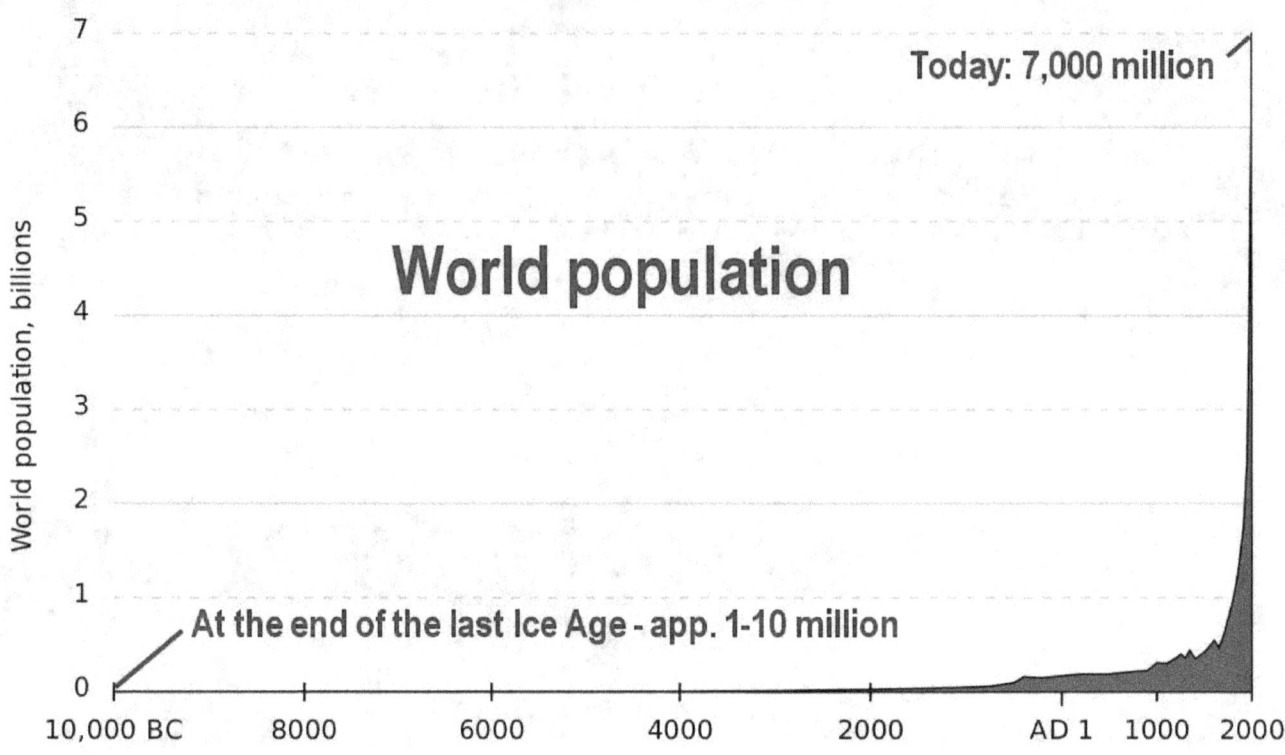

In spite of the millstones that have been hung around humanity's neck during the hellish period of its history, of the kingdoms of empire, humanity has forged ahead to ever-greater levels of self-development, in science, technology, and humanity. The proof for this is amazingly evident in the 'pudding' of increasing population density that has been achieved.

The proof of the 'pudding' is unmistakable. Until scientific thinking began to dawn in the world, at around 300 BC, the world population had remained almost stagnant. Then with the scientific methods of Socrates and Plato setting a higher stage for humanity, a new wind began to blow that opened the horizon of the mind.

the sky is no limit" or "there are no limits."

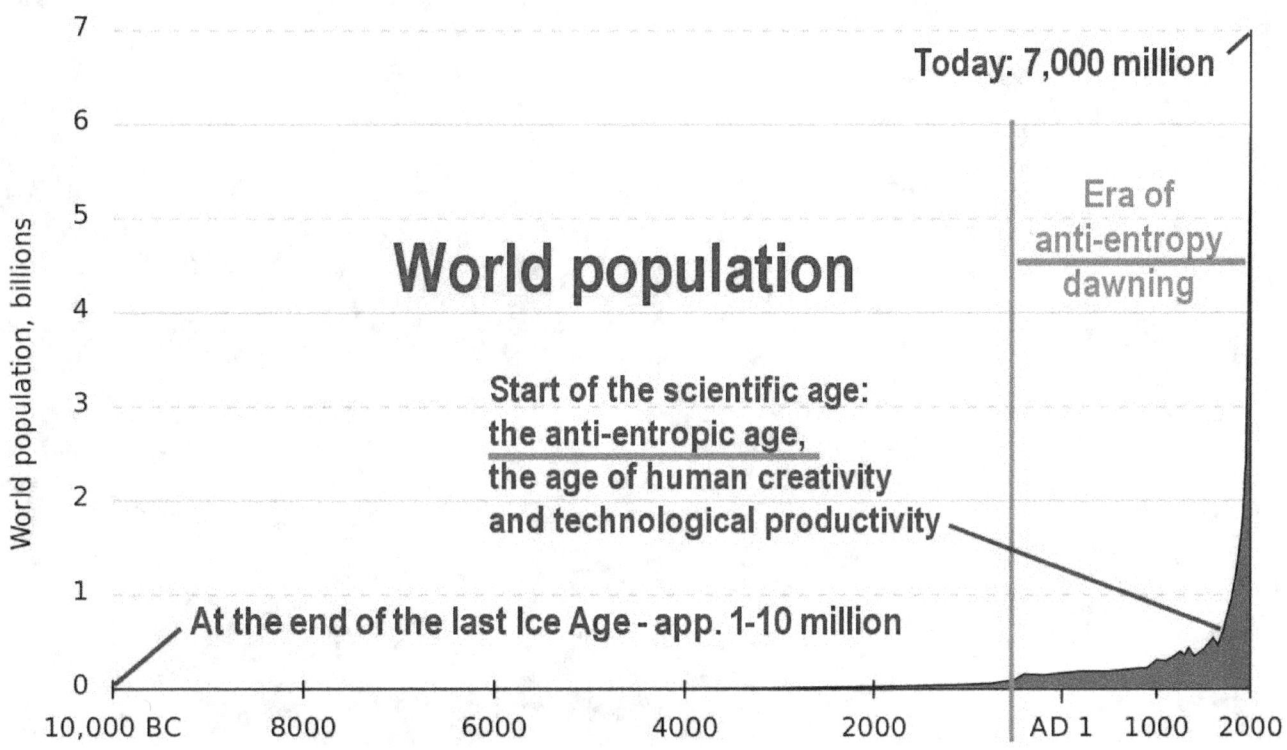

Suddenly, with small-minded thinking becoming put aside, a dramatic increase in economic power began to develop. The increase was slow at first, choked by the devastating effects of the continuing empires of the world. The chokehold was broken, however, by the new breakthrough in science that unfolded with the work of Johannes Kepler, who started a new freedom in thinking and in humanity's self-perception.

As the result, society became more productive, more innovative, more creative, as if the watchword was spoken, "the sky is no limit" or "there are no limits."

# With increased industrialization

With increased industrialization, and advances in farming, mechanization, transportation, and so on, in an environment of increased energy use, it became possible for evermore people to support themselves on the lands of the Earth that in primitive times supported just a few.

The sharp increase in the human world population to the 7 billion level in our time, was obviously not the result of improved breeding habits, but does simply reflect the anti-entropic power of humanity becoming increasingly recognized, with which society creates itself new resources for living with advanced technologies, and so on, all provided by the anti-entropic human intellect.

# Population increase mirrors anti-entropic economics

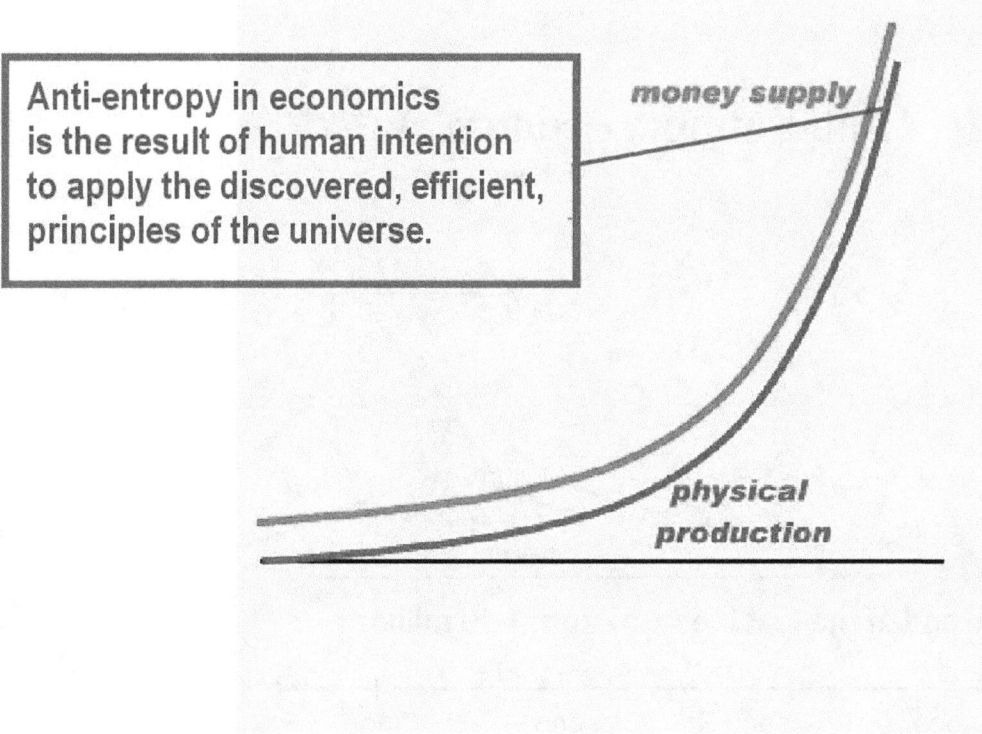

The worldwide achieved population increase mirrors the natural dynamics of anti-entropic economics.

The graph shown here illustrates merely the principle of the dynamics. The principle is addressed in Segment 2 of the video series on the Big Bang entropy theory.

## The rate of increase that can be achieved

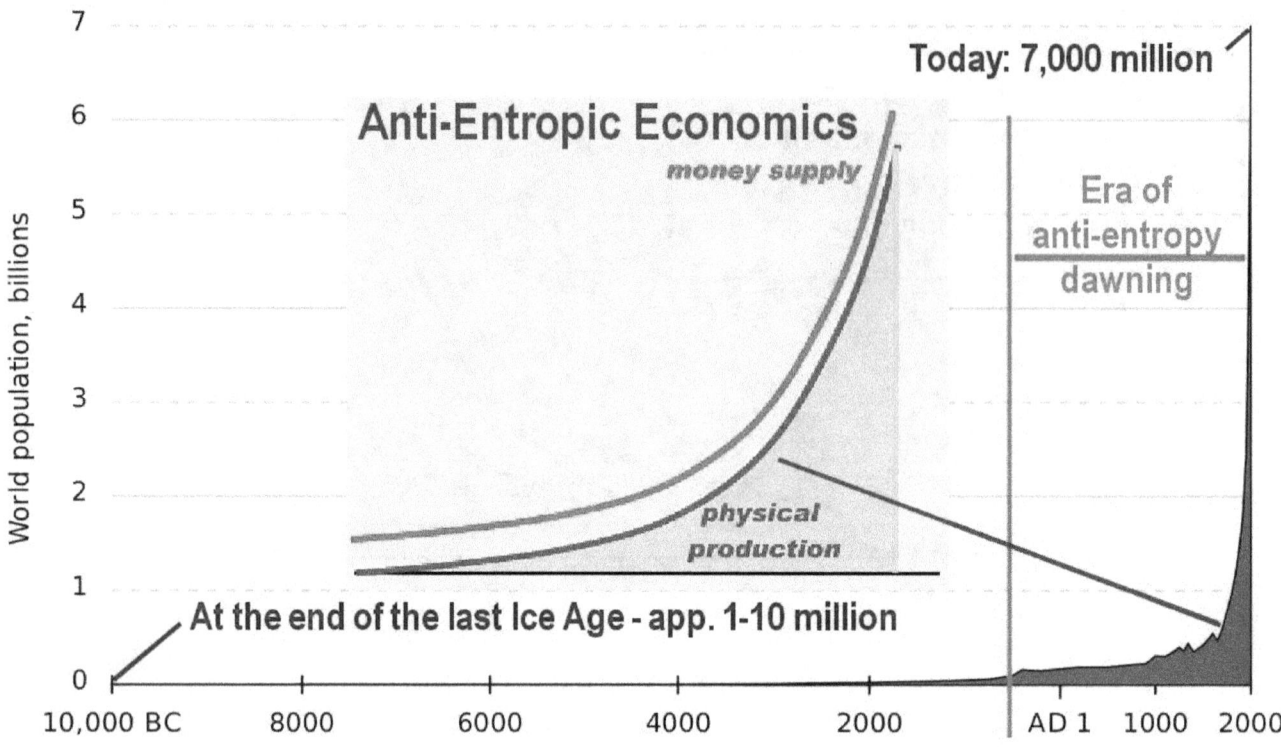

The rate of increase that can be achieved, of the expression of the dynamics, is only limited by the limits society is placing on the human spirit in its surging ahead.

# Floating bridges across the tropical seas

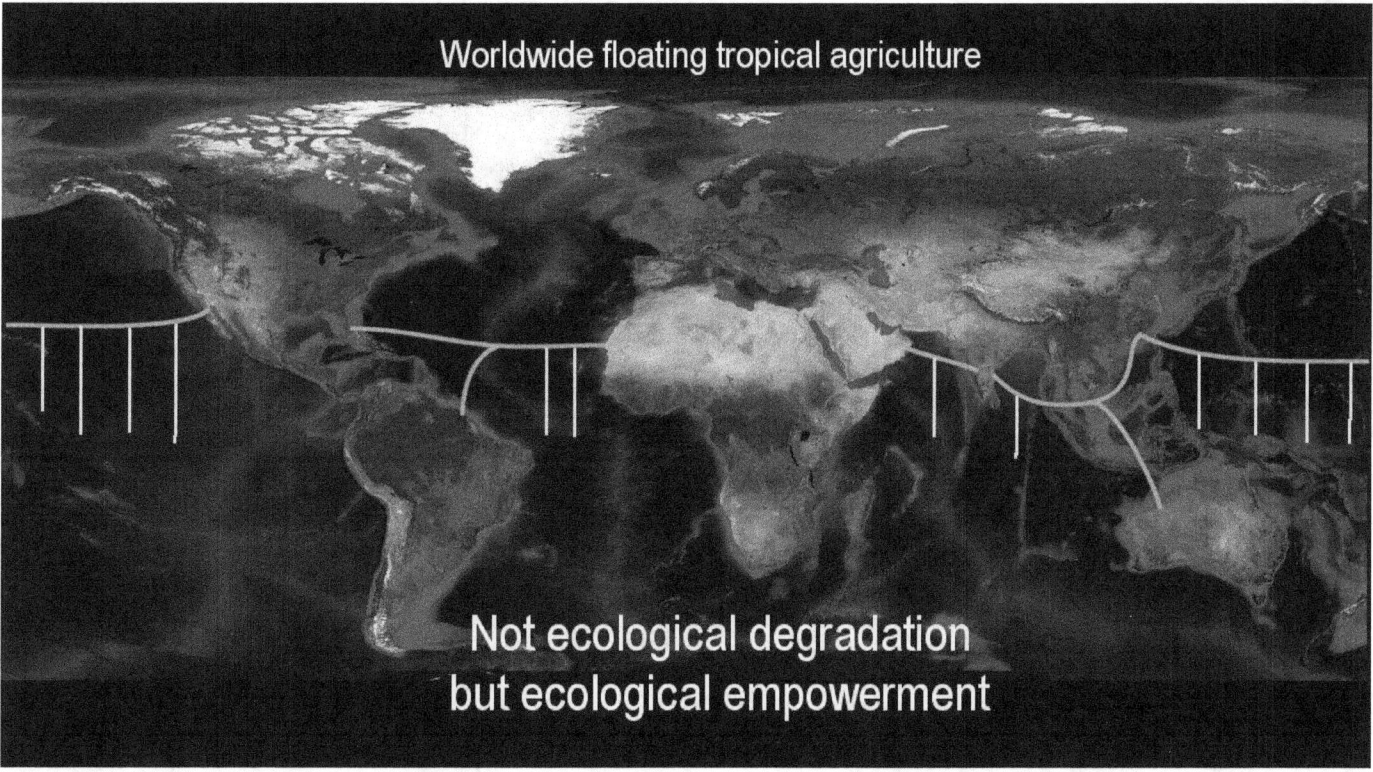

What would prevent humanity today from laying down floating bridges across the tropical seas that connect up with floating agriculture that will take over the food production for the world when the recurring Ice Age disables agriculture outside the tropics, potentially in the 2050s timeframe.

We have the materials for it in the form of basalt, and the energy resources for it in the form of thorium nuclear fission. We have the technology already in use on a small scale. Once we can get our spirit roused to do this, this gigantic seeming project will become a small thing.

# The 6000 new cities for a million people each

We would likely place most of the 6000 new cities for a million people each, which we will need, afloat onto the seas, together with their agriculture, in preparation for relocating the nations out of the northern areas that become uninhabitable in an Ice Age environment, and we would create the cities for one-another for free. With automated, high-temperature industrial production, the mass-production of quality housing will become so easy that cost-free living will become the new basic infrastructure for humanity meeting its human needs in the most-efficient manner possible, as an open door for further advances in scientific and cultural achievements. All of this becomes possible when we lay aside the entropic, small-minded, mode of self-perception, to an open-ended perception.

## To break down the barrier in the mind

Our greatest need therefore, is to break down the barrier in the mind that would prevent us from meeting our potential as human beings. The barrier of small-mindedness is already killing far too many people with poverty, even in the richest of the rich nations. With the increasing drought conditions, as a fringe effect towards the coming Ice Age, the USA is fast running out of freshwater resources. The shortage could have been prevented.

## Drought conditions can be offset

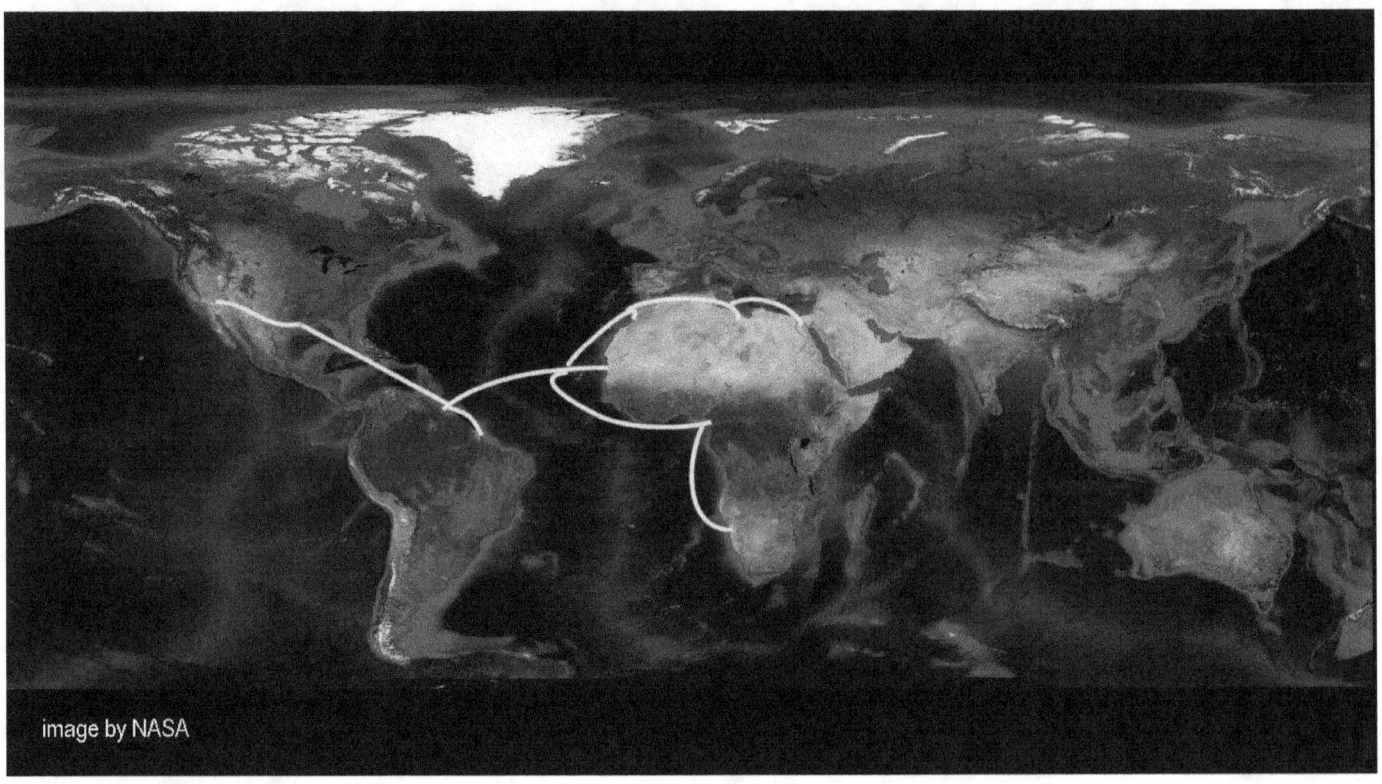
image by NASA

Drought conditions can be offset by simply redirecting some of the outflow of the great tropical rivers, such as the Amazon River, to the dry areas of the world via a network of floating arteries made of woven basalt.

## Desalination of ocean water provides an endless resource

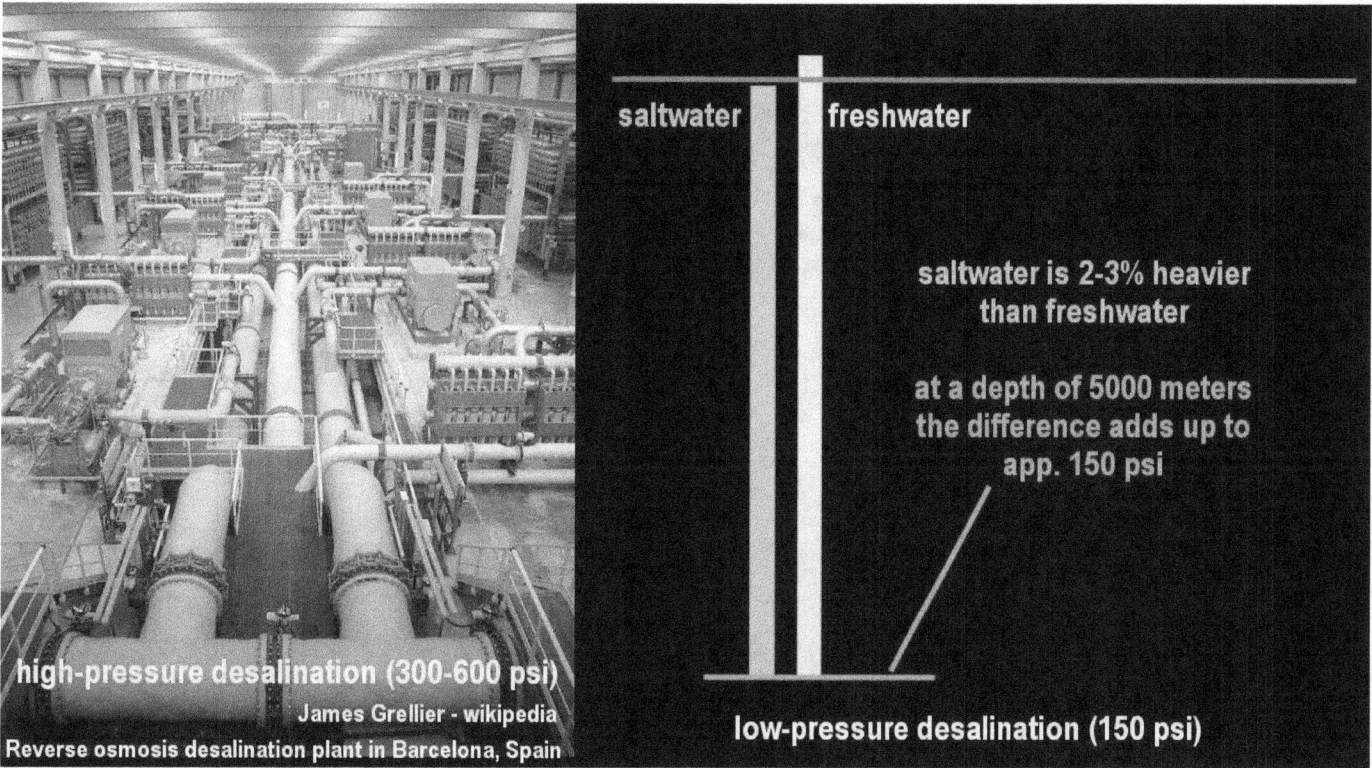

Actually, we don't need to rely on rivers at all, in order to meet our freshwater needs. Desalination of ocean water provides an endless resource. All we need to do is utilize the weight differential between freshwater and salt water, to self-power deep ocean reverse osmosis desalination systems, to cause rivers of freshwater to flow out of the oceans, independent of weather conditions. The technology already exists. The deep-ocean application hasn't been implemented yet, because small-minded, entropic thinking that accepts the collapse of human living as normal, has prevented the application of the humanist power we already have at hand.

## Recognized already during the Kennedy era

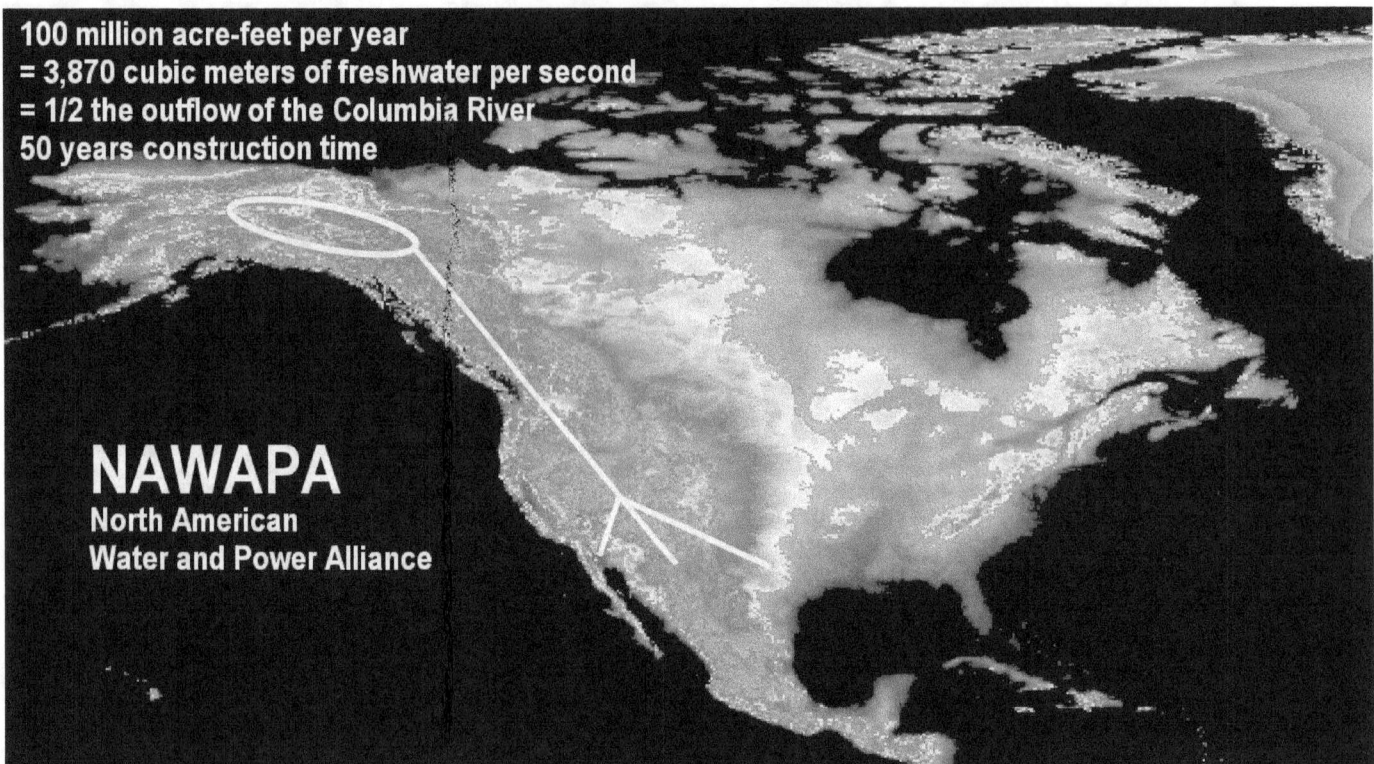

It was recognized already during the Kennedy era that the south-western dry areas of the USA could be made far more productive with increased volumes of freshwater. For this reason a project was devised to divert portions of the large northern rivers to the south for increased agricultural production. The giant NAWAPA project was never even started. Now, with the drought setting in, the Southwest is in a water crisis that is decimating agricultural production in the region, instead of increasing it.

# Deep ocean reverse osmosis desalination

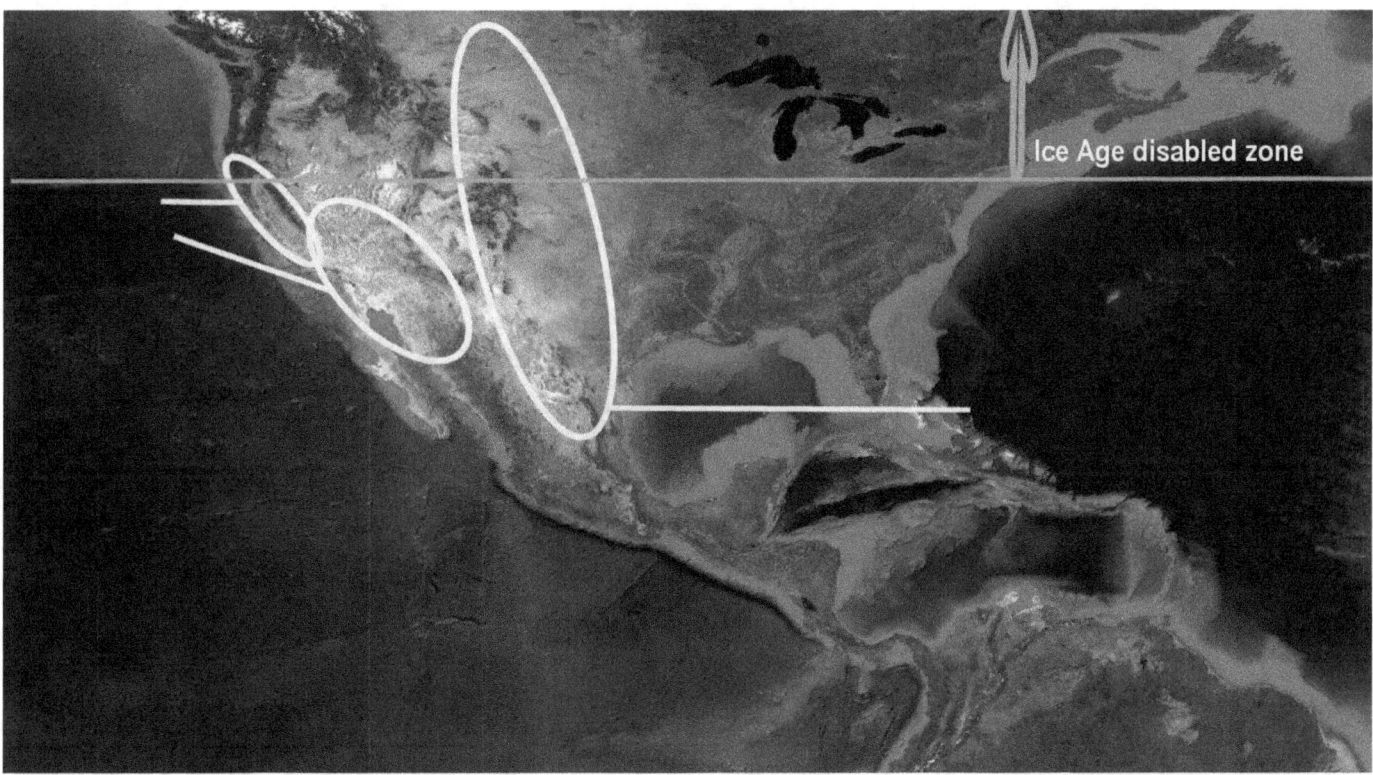

While it would take a 50-year construction effort to divert water from Alaska, deep ocean reverse osmosis desalination can be built relatively easily. The deep oceans exist. The volume of freshwater that can be produced depends only on the size of the infrastructures that we built to meet our needs. And those needs will increase, dramatically, especially during the coming Ice Age environment when rain becomes scarce. Desalination becomes essential then, and not only for agriculture.

## Deep ocean reverse osmosis desalination

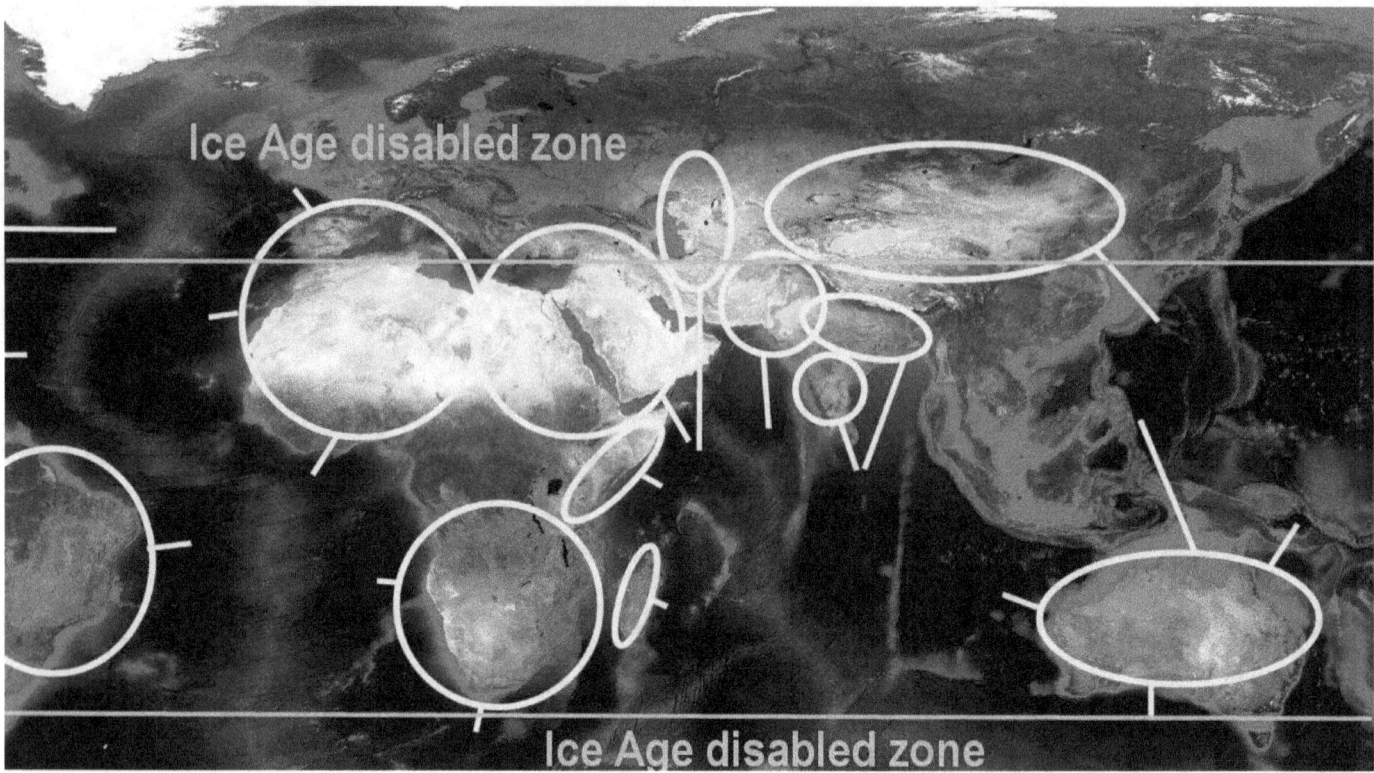

The entire world will find itself in the same situation. It has the same opportunity for unlimited freshwater production from deep ocean reverse osmosis desalination that America has available to it. Deep oceans are nearby everywhere.

# India can never suffer water shortages

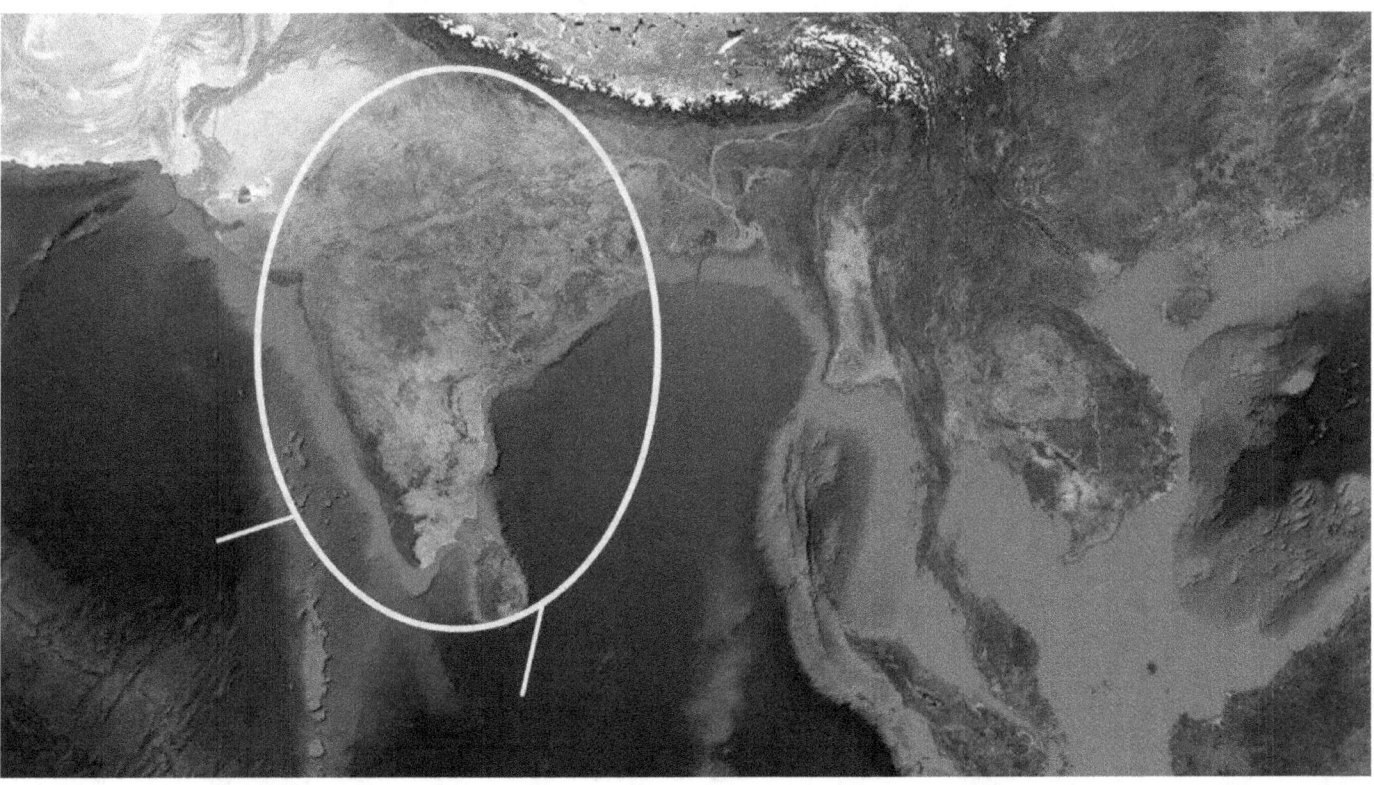

India can never suffer water shortages if the infrastructures are being built. Those will become critically essential for the coming Ice Age when the rain becomes scarce. This means that the infrastructures should be started now, including the inland distribution systems. Large infrastructures are not built quickly, especially when the industries for building them need to be created first. It is tempting to say, that these gigantic efforts are not possible, but in saying this we would be denying ourselves. We would deny us as an anti-entropic species. With this denial we would be laying ourselves down to die.

# The relocation of entire nations becomes essential

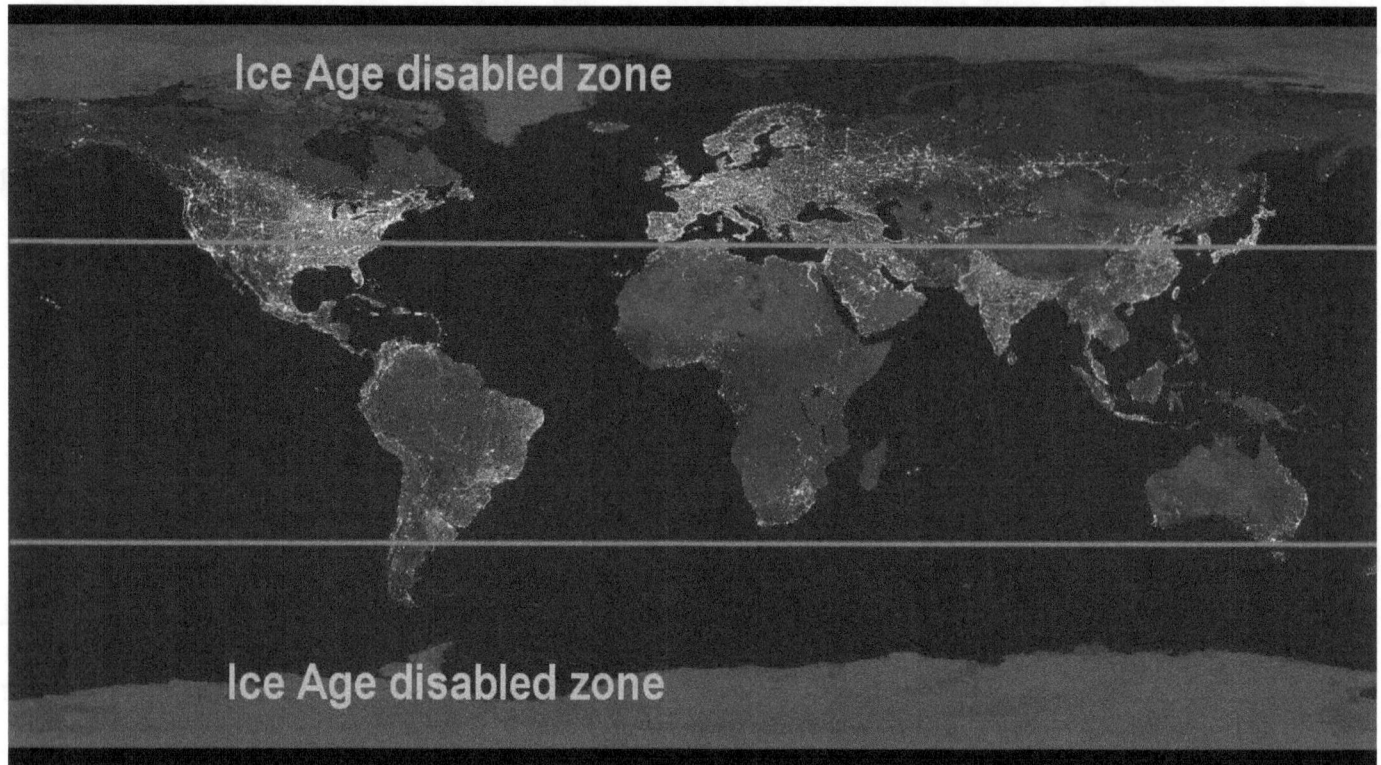

The tropical world, in which the future of humanity will have to be located in 30 years when the Ice Age begins anew, is presently vastly underdeveloped, while the highly developed regions are located in the zones that become uninhabitable. The relocation of entire nations becomes thereby essential. This means that enormously large projects will become common, because it is unthinkable that humanity will simply lay itself down and commit suicide by default. This won't happen.

## We will built the 6000 new cities

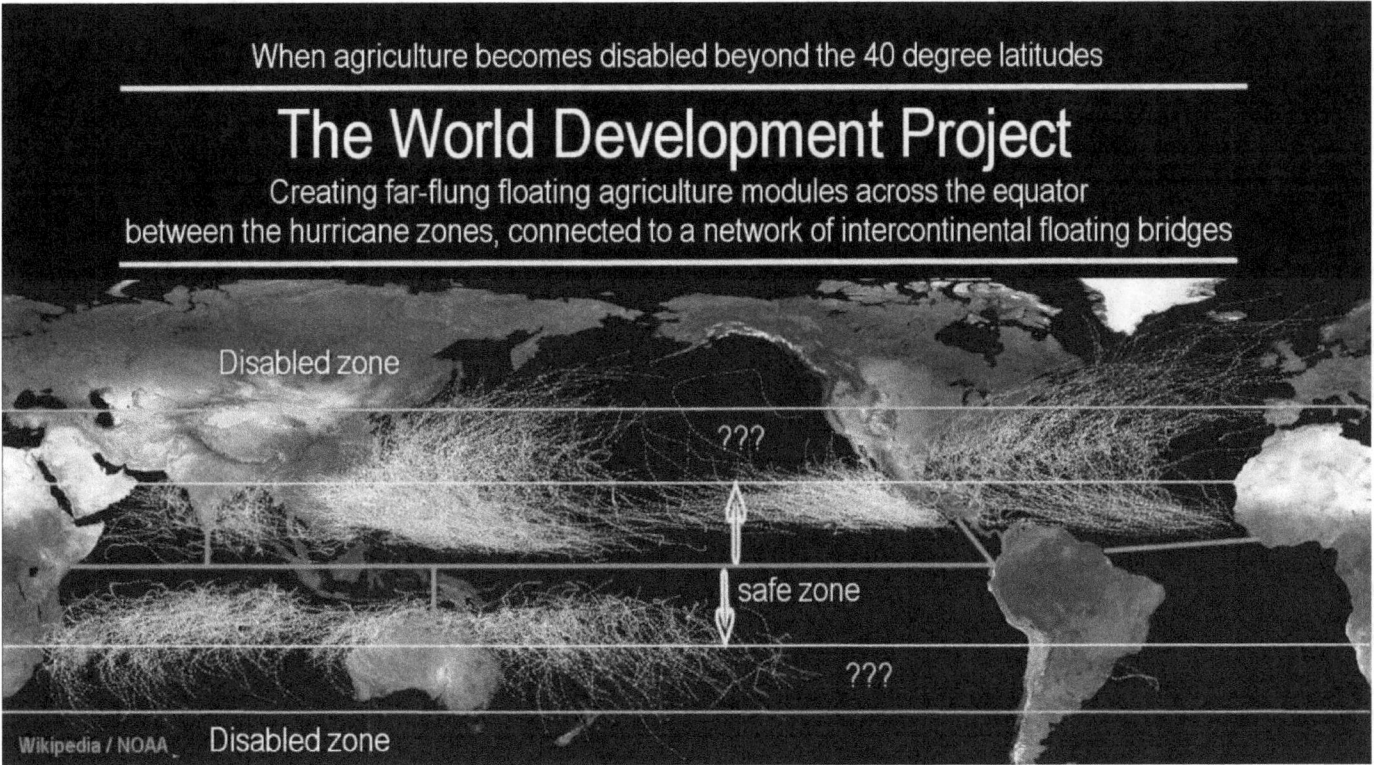

Sure, some decisive efforts will need to be made to get us all out of the present slump of small-minded thinking that keeps us tied to a decaying landscape where impotence and entropy rules, especially economic entropy that invites stealing, where we say this is too hard to do, who will pay for it all, so that nothing is presently being built. However, the prospect isn't pleasant either, that when the Ice Age begins in the 2050s and devastates agriculture outside the tropics, that most of humanity will starve to death, because people don't live long without food. Consequently, we will stir our stumps and get the job done, to meet our future needs. This means that we will built the 6000 new cities for a million people each and the millions of acres of floating agriculture that are required, because this is the human thing to do, and an exceedingly exciting thing to do as well. Most likely, we haven't seen anything yet in terms of what we can accomplish as human beings with our profound intellect and creative capacity. This will set the stage for the future.

## Whether we survive the Ice Age Challenge

Whether we survive the Ice Age Challenge with vast new infrastructures in place in the tropics in preparation for a potential Ice Age Renaissance world, or whether we will fail ourselves and die of starvation as a consequence, depends exclusively on the recognition of ourselves as an anti-entropic species that inspires us to accomplish the needed tasks.

## We have ample of proof to our credit

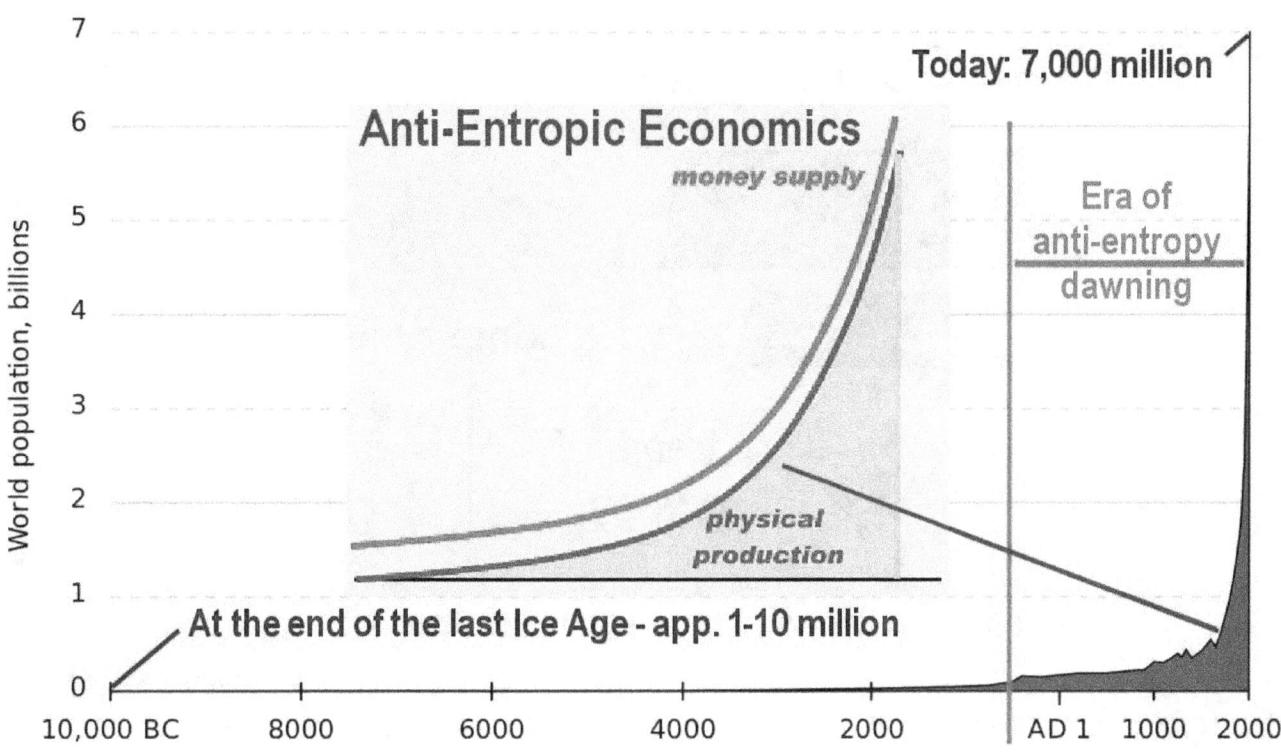

I would like to suggest that we will not fail ourselves, because we have already demonstrated to ourselves that we are vastly more capable than we presently give ourselves credit for, for which we have ample of proof to our credit.

## Entropic factions in politics and philosophy

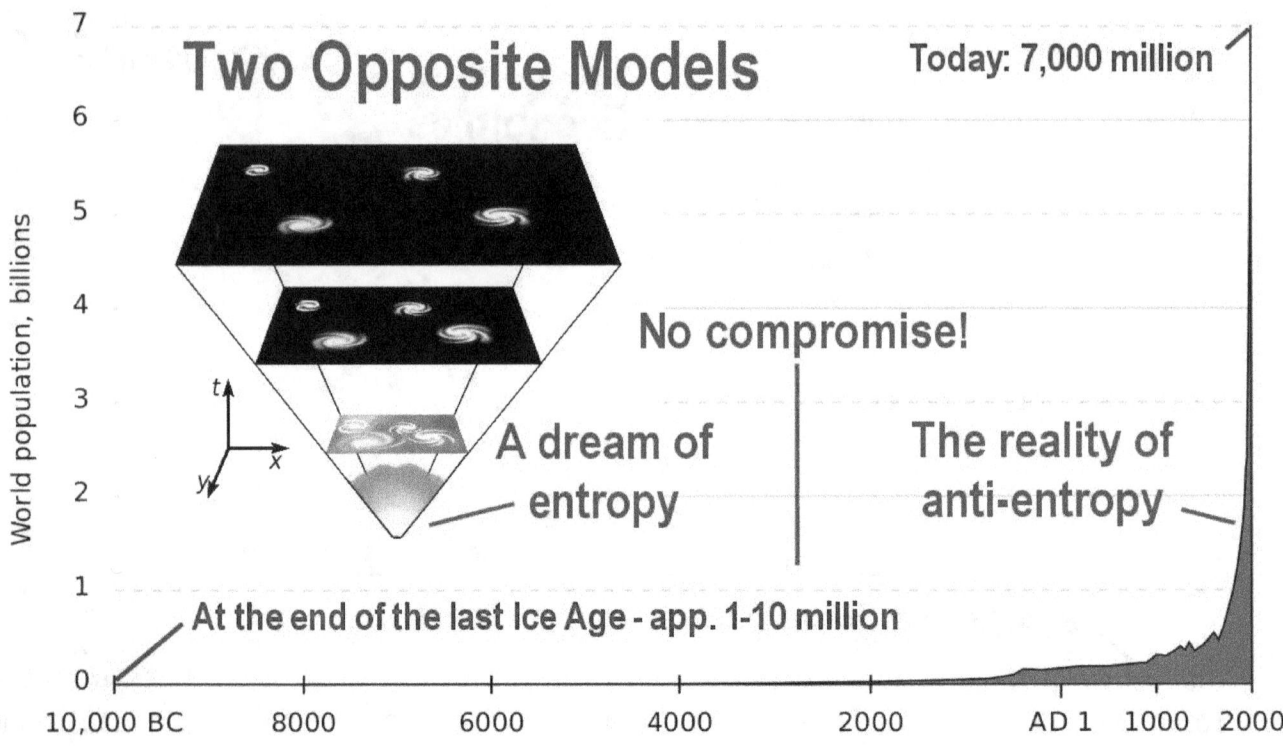

While entropic factions in politics and philosophy see the amazing proof of the human anti-entropy with fear, and say that humanity is overgrazing the land that will collapse the biosphere and thereby diminish the resources, they speak from ignorance and self-dilution. Their fear is false, built of false historic philosophies that are as false as the Big Bang theory is false that appears to be intentionally entropic.

The reality is that nothing is winding down anywhere in the universe, nor is the potential of humanity winding down. Our open-ended development potential has been demonstrated as truth, and the truth does not diminish. In fact we have just begun to peck open the shell of our infancy, reaching for infinity.

## As the truth is being experienced, we begin to fly high

In discovering our truth, by the discovery of the creative principles of the universe, we discover ourselves as the supreme being on Earth in the living image of God or the Universe.

Then, as the truth is being experienced, we begin to fly high above the dust of entropy and find that our world has become cleansed of all of its pesky, numerous, small-minded illusions, which thereby has become transformed into fruitful fields and gardens of beauty and happiness. And for that, we have the means already at hand.

# More Illustrated Science Books by Rolf A. F. Witzsche

www.ingramcontent.com/pod-product-compliance
Lightning Source LLC
Chambersburg PA
CBHW080955170526
45158CB00010B/2811